Oxford Lecture Series in
Mathematics and its Applications

Series Editors
John Ball Dominic Welsh

OXFORD LECTURE SERIES
IN MATHEMATICS AND ITS APPLICATIONS

Books in the series

1. J. C. Baez (ed.): *Knots and quantum gravity*
2. I. Fonseca and W. Gangbo: *Degree theory in analysis and applications*
3. P. L. Lions: *Mathematical topics in fluid mechanics, Vol. 1: Incompressible models*
4. J. E. Beasley (ed.): *Advances in linear and integer programming*
5. L. W. Beineke and R.J. Wilson (eds): *Graph connections: Relationships between graph theory and other areas of mathematics*
6. I. Anderson: *Combinatorial designs and tournaments*
7. G. David and S.W. Semmes: *Fractured fractals and broken dreams*
8. Oliver Pretzel: *Codes and algebraic curves*
9. M. Karpinski and W. Rytter: *Fast parallel algorithms for graph matching problems*
10. P. L. Lions: *Mathematical topics in fluid mechanics, Vol. 2: Compressible models*
11. W. T. Tutte: *Graph theory as I have known it*
12. Andrea Braides and Anneliese Defranceschi: *Homogenization of multiple integrals*
13. Thierry Cazenave and Alain Haraux: *An introduction to semilinear evolution equations*
14. J. Y. Chemin: *Perfect incompressible fluids*
15. Giuseppe Buttazzo, Mariano Giaquinta and Stefan Hildebrandt: *One-dimensional variational problems: an introduction*
16. Alexander I. Bobenko and Ruedi Seiler: *Discrete integrable geometry and physics*
17. Doina Cioranescu and Patrizia Donato: *An introduction to homogenization*
18. E. J. Janse van Rensburg: *The statistical mechanics of interacting walks, polygons, animals and vesicles*
19. S. Kuksin: *Hamiltonian partial differential equations*
20. Alberto Bressan: *Hyperbolic systems of conservation laws: the one-dimensional Cauchy problem*
21. B. Perthame: *Kinetic formulation of conservation laws*
22. A. Braides: *Gamma-convergence for beginners*
23. Robert Leese and Stephen Hurley: *Methods and Algorithms for Radio Channel Assignment*
24. Charles Semple and Mike Steel: *Phylogenetics*
25. Luigi Ambrosio and Paolo Tilli: *Topics on Analysis in Metric Spaces*
26. Eduard Feireisl: *Dynamics of Viscous Compressible Fluids*
27. Antonín Novotný and Ivan Straškraba: *Introduction to the Mathematical Theory of Compressible Flow*
28. Pavol Hell and Jarik Nesetril: *Graphs and Homomorphisms*
29. Pavel Etingof and Frederic Latour: *The dynamical Yang-Baxter equation, representation theory, and quantum integrable systems*
30. Jorge Ramirez Alfonsin: *The Diophantine Frobenius Problem*
31. Rolf Niedermeier: *Invitation to Fixed Parameter Algorithms*
32. Jean-Yves Chemin, Benoit Desjardins, Isabelle Gallagher and Emmanuel Grenier: Mathematical Geophysics: *An introduction to rotating fluids and the Navier-Stokes equations*
33. Juan Luis Vázquez: *Smoothing and Decay Estimates for Nonlinear Diffusion Equations*

Smoothing and Decay Estimates for Nonlinear Diffusion Equations

Equations of Porous Medium Type

Juan Luis Vázquez

Dpto. de Matemáticas
Univ. Autónoma de Madrid
29049 Madrid, Spain

OXFORD
UNIVERSITY PRESS

OXFORD
UNIVERSITY PRESS

Great Clarendon Street, Oxford OX2 6DP
United Kingdom

Oxford University Press is a department of the University of Oxford. It furthers the University's objective of excellence in research, scholarship, and education by publishing worldwide. Oxford is a registered trade mark of Oxford University Press in the UK and in certain other countries

First published 2006
Reprinted 2013

Published in the United States of America by Oxford University Press
198 Madison Avenue, New York, NY 10016, United States of America

British Library Cataloguing in Publication Data
Data available

Library of Congress Cataloging in Publication Data
Data available

ISBN 978-0-19-920297-3

Preface

This text is concerned with quantitative aspects of the theory of nonlinear diffusion equations. These equations can be seen as nonlinear variations of the classical heat equation, the well-known paradigm to explain diffusion, and appear as mathematical models in different branches of physics, chemistry, biology and engineering. They are also relevant in differential geometry and relativistic physics. Much of the modern theory of such equations is based on estimates and functional analysis. Indeed, nonlinear functional analysis is a quite active branch of mathematics, and a large part of its activity is aimed at providing tools for solving the equations originated in scientific disciplines like the above-mentioned.

We concentrate on a class of equations with nonlinearities of power type that lead to degenerate or singular parabolicity and we gather these collectively under the name 'equations of porous medium type'. Particular cases are the porous medium equation (PME), the fast diffusion equation (FDE) and the evolution p-Laplacian equation. These equations have a wide number of applications, ranging from plasma physics to filtration in porous media, thin films, Riemannian geometry, and many others; and they have at the same time served as a testing ground for the development of new methods of analytical investigation, since they offer a variety of surprising phenomena that strongly deviate from the heat equation standard. Among those phenomena we count free boundaries, limited regularity, mass loss, and extinction or quenching, to quote a few.

The aim of the present work is obtaining sharp a priori estimates and decay rates for general classes of solutions of those equations in terms of estimates of particular problems. The estimates will be building blocks in understanding the qualitative theory, and the decay rates should pave the way to the fine study of asymptotics. Basic tools are results of symmetrization and mass concentration comparison, combined with scaling properties; all of this reduces the problem to getting a detailed knowledge of special solutions using worst-case strategies. The functional setting consists of Lebesgue and Marcinkiewicz spaces, and our final aim is to get a deeper knowledge of the evolution semigroup generated by the equation. We obtain optimal estimates with best constants. Many technically relevant questions are presented and analysed in detail, like the question of strong smoothing effects versus weak smoothing effects. The end result combines a number of properties that extend the linear parabolic theory with an array of peculiar phenomena. As a summary, a systematic picture of the most relevant phenomena is obtained for the equations under study, including time decay, smoothing, extinction in finite time, and delayed regularity.

Being based on estimates, this is essentially a book about mathematical inequalities and their impact on the theory. A classic in that respect is no doubt the treatise

'Inequalities' by G. Hardy, J. E. Littlewood and G. Pólya [HLP64]. Another source of motivation is the famous line of inequalities known collectively as Sobolev inequalities, that have permeated the study of nonlinear PDEs since the middle of the twentieth century. We recall that in mathematics an inequality is simply a statement about the relative size or order of two objects. Our inequalities determine or control the behaviour of nonlinear diffusion semigroups in terms of data and parameters. That sums up our game in simple terms.

The present text contains results taken from papers of the author and collaborators on the theory of nonlinear diffusion, and also the main progress due to other authors. Together with the monograph [V06] in which we develop the mathematical theory of the PME, the surveys [Va03, Va04] on asymptotic behaviour, and the text co-authored with V. Galaktionov on a dynamical systems approach to nonlinear PDE evolution problems [GV03], it represents the effort of the author to present to a wide audience a substantial part of the work involving the PME/FDE that has been developed in the last few decades, and as a support for the work that continues to be done nowadays in new directions. The book contains a fair amount of new results and open problems; actually, we feel that further ideas and understanding are still needed in this area, and even more in its many interactions with other subjects in the wide world of nonlinear PDEs.

Acknowledgments

This text is the result of many years of thinking on the topics of nonlinear semigroups, bounds and asymptotics. It is a pleasure to mention some of the people who made possible this particular journey through the kingdom of Nonlinear Diffusion.

My interest in the topic started decades ago under the influence of the late Philippe Bénilan who always thought about nonlinear diffusion problems in functional terms; his mind was busy with functional bounds and semigroups, and he made some of the basic contributions on which the text is built; in that connection and time, Laurent Véron also had a strong influence trough his classical paper [Ve79]. Next come two of the main techniques: I learnt symmetrization from Giorgio Talenti and the art of self-similarity from Shoshana Kamin, Bert Peletier, and Grisha Barenblatt. The books of the latter are a continuous source of inspiration and enjoyment and an open window into the Russian school of mathematics.

Many of the topics reported here originate from works with collaborators, too numerous to quote; I would like to single out the inspiration I received for this research from Don Aronson and Luis Caffarelli, with whom I spent happy periods in the USA and wrote some of my best contributions. Later, I was strongly influenced by Victor Galaktionov, who loves asymptotics. I would also like to thank Haim Brezis for his continuous encouragement of my mathematics; besides, he produced the first smoothing effects applicable to a large class of nonlinear evolution equations including the porous medium equation; he also pioneered the study of Radon measures as data, and he wrote with Avner Friedman a very influential paper on non-existence for fast diffusion, a favourite topic for me. The presentation of the geometric aspects of fast diffusion owes much to conversations with Panagiota Daskalopoulos. Work on the p-Laplacian was shared with Lucio Boccardo and Thierry Gallouët.

Finally, this work would not have been possible without the scientific contributions and personal help of my former students Ana Rodriguez, Arturo de Pablo, Fernando Quirós, Guillermo Reyes, Juan Ramón Esteban, Manuela Chaves, Omar Gil and Raúl Ferreira, to whom I would like to add Emmanuel Chasseigne and Matteo Bonforte.

The final index lists the main concepts and the names of the authors of the results that have been most influential on the author in writing this text, as mentioned in the different chapters. The author apologizes for undue omissions in the list and the citations.

This work was partially supported by Spanish Project BFM2002-04572-C02. Part of this work was performed while visiting ICES, University of Texas at Austin, as Oden Fellow in 2004.

Madrid, 2005

Key words. Nonlinear parabolic equations, smothing effect, decay of solutions, nonlinear semigroups, scaling, symmetrization, concentration, Marcinkiewicz spaces extinction.

AMS Subject Classification. 35B05, 35B40, 35K55, 35K65, 47H20.

Contents

Appendices

Introduction

A modern approach to the existence and regularity theory of partial differential equations relies on obtaining suitable a priori estimates in terms of the information available on the data, typically in the form of norms in appropriate functional spaces. Following such ideas, this book is devoted to obtaining basic estimates for some particular nonlinear parabolic equations and to derive consequences about qualitative and quantitative aspects of the theory. In pursuing this aim, a major role is given to the scaling properties and the existence of suitable self-similar solutions; results of symmetrization and mass concentration comparison also play a prominent role; finally, a strategy of looking for the worst case completes the picture.

Using this machinery, we can derive a complete set of a priori estimates in Lebesgue and Marcinkiewicz spaces for the two model equations of nonlinear diffusion theory:

$$u_t = \Delta u^m, \quad u_t = \nabla \cdot (|\nabla u|^{p-2}\nabla u), \tag{1}$$

which we study for the different values of the exponents m and p and space dimension $n \geq 1$. Both equations are popular models in that area, with a number of applications in physics and other sciences, and with a very rich mathematical theory. We will give preference in the presentation to the first equation, which is known in the literature as the porous medium equation (shortly, PME) when $m > 1$, and the fast diffusion equation (FDE) when $m < 1$. The classical heat equation (HE) is included as the case $m = 1$, but contrary to the standard approach for the latter equation, our approach here is heavily nonlinear in methods and results. The bulk of the book is devoted to the analysis of smoothing estimates and the related topics that arise for the porous medium equation and its relative, the fast diffusion equation.

The second equation mentioned above is usually called the p-Laplacian evolution equation, and is the best known of a series of models of nonlinear parabolic equations called "gradient-dependent diffusion equations". Our interest in including it here, even if with a lower level of attention, serves the purpose of showing that the systematic study of the PME / FDE is applicable to the p-Laplacian equation, and also to a number of usual models that appear all the time in the theory and applications of nonlinear diffusion, and more generally, in reaction diffusion.

The tools

Let us review the technical tools: the topics of symmetrization and mass concentration comparison relevant for our study have been explained in detail in the paper [Va04b], which was focused on a more general model, the so-called filtration equation,

$$u_t = \Delta\varphi(u) + f, \tag{2}$$

where usually φ is a C^1 real function with positive derivative; in degenerate cases φ' is only non-negative; the equation may also include singularities, i.e., values where φ is not C^1 and $\varphi'(u) = \infty$. Equation (2) includes as a particular case our main model, $u_t = \Delta u^m$, as well as other popular nonlinear diffusion models, like the Stefan equation, where $\varphi(u) = (u-1)_+$. Our results in this paper have corollaries for such equations, though we will not develop them. Many nonlinear diffusion variants can also be treated by similar methods, like the already mentioned p-Laplacian equation, and we will give details of this extension.

On the other hand, the model equations we consider have the extra property of being invariant under a large group of scaling transformations, with at least two free parameters (see formula (1.22) below). Indeed, the property of scale invariance is the connecting link between the two otherwise quite different parabolic equations (1). Moreover, both equations can be combined into a more complicated model, so-called doubly nonlinear equation, to which the methods apply. A striking consequence of scale invariance is the existence of self-similar solutions whose behaviour and properties can be described in great detail. Special cases of such solutions will serve as model examples in our worst-case strategy to obtain estimates. They are the main stars of our show.

Let us point out that obtaining particular solutions of PDEs does not seem a fundamental problem in the most theoretical approach to the subject, but reality is different under such a deceptive cover: self-similarity, separation of variables, the Bäcklund transformation, the method of characteristics, Green's function, integral transforms, and other methods allow the applied mathematician to gain the insight that serves as a corner-stone for the general treatment. This approach will lead our steps in what follows.

The goal. Smoothing

The present work uses the approach we have outlined to derive a complete set of a priori estimates that should play an important role in the qualitative theory of the equations. Let us present the motivation for this application, starting from the classical heat equation, $u_t = \Delta u$. This equation has a remarkable property, called *smoothing*, whereby solutions with (initial and boundary) data in suitable functional spaces are actually C^∞ smooth functions in the interior of the domain of definition. This is a result easily obtained in the case of the Cauchy problem posed in the whole space by using the standard representation with the Gaussian kernel. It was soon observed that the property is shared by a whole class of so-called linear parabolic equations with variable, even non-smooth coefficients, being remarkable in this context the work of J. Nash [Na58, KN02], while the elliptic counterparts bear the names of E. De Giorgi [DG57] and J. Moser [Mo61]. The results were later extended to wide classes of nonlinear equations under suitable structural assumptions, the main ones being uniform parabolicity (or ellipticity) and smoothness of the coefficients. Uniform parabolicity will be missing in this text.

When performing the regularity proofs in more general contexts, one is led to proceed by steps: first, data in general spaces, like $u_0 \in L^p$ spaces, produce solutions which are bounded, $u \in L^\infty$, and this is the starting point of the 'improvement process'; in a second step, bounded solutions are shown to be continuous, usually Hölder continuous, by means of a priori interior bounds in terms of the proven bounds for the solution; third, an iterative argument allows us to obtain estimates for first derivatives and then derivatives of all orders. When dealing with increasingly wider classes of equations, mainly nonlinear equations or equations with bad coefficients, the latter steps may or may not be true (a phenomenon called partial regularity), and the first step receives special attention, being the most general. This is why we will accept some usual terminology in nonlinear PDE analysis and give the (rather incorrect) name of *smoothing effect* to the property which could be better termed 'function space improvement' and says: data in a space like $L^p(\mathbf{R}^n)$ produce solutions that live for $t > 0$ in a space $L^q(\mathbf{R}^n)$ with $q > p$, hopefully $L^\infty(\mathbf{R}^n)$. Describing when the aforementioned equations do or do not possess such a smoothing effect, what are the quantitative estimates behind that property, or what happens otherwise is the main purpose of this text.

Similar estimates are well known in the theory of linear equations and semigroups. Semigroups that send initial data in $L^1(\Omega, dx)$ into orbits in $L^\infty(\Omega, dx)$ are called ultracontractive, see [Dv89, Chapter 2], and this is the type of property of concern for us. But the reader should notice that while for linear equations any boundedness estimate is equivalent to a stability result (i.e., control of differences of solutions in terms of differences of data), this is not at all the case for nonlinear semigroups. Since the equation $u_t = \Delta u^m$ is nonlinear for $m \neq 1$, the estimates that give boundedness of the PME/FDE flow do not necessarily imply any kind of contractivity or stability, and this is a main source of divergence in the corresponding theory. We will return to this issue in Section 4.2.

Semigroups, bounds, asymptotic rates, and patterns

There are many sides in this task under the flag of nonlinear evolution. Thus, from the point of view of functional analysis, the whole question of smoothing effects can be seen as a rather basic part of the study of the evolution semigroups generated by the equations, as was pointed out in the pioneering works of Bénilan [Be72, Be76] and Véron [Ve79]. As we will see, it gives in some cases the information needed to properly define the semigroup in a suitable domain; in any case, it shows how the semigroup behaves in time.

The text also contributes to the topic of asymptotic behaviour. Given an equation E, the idea is to associate to every data, in our case initial data u_0, a set of so-called 'asymptotic data' that allow us to reconstruct the long-time behaviour of the solution generated by E from u_0. In the standard application the solution of a nonlinear diffusion process exists globally in time and goes to zero in a more or less uniform way. The asymptotic data must be simple to calculate and must allow us to reconstruct the approximate behaviour of $u(x,t)$ for large times (so-called *intermediate asymptotics*); they take the form of *asymptotic rates* (which are usually powers of time), as well as the spatial pattern, usually called the *asymptotic profile*. We have contributed other studies on the asymptotic behaviour of the PME [Va03, Va04]. Some of the main new contributions of our text lie in the realm of fast diffusion and are related to extinction phenomena. By this we mean that for some evolution equations and data the solution disappears completely at a certain time $T > 0$ in the sense that $u(x,t) \to 0$ at all points $x \in \mathbf{R}^n$ as $t \nearrow T$.

This work aims at giving the reader a sound knowledge of the behaviour of the solutions of the semigroups generated by the two families of equations (1), specially the first one. The idea of comparison with the well-known heat equation and its Gaussian kernel is always present. The similarities and striking differences will be outlined. Our main characters will be the self-similar functions that play the role of the Gaussian kernel in the different contexts, and the plot will almost always revolve around them.

Contents and distribution

The analysis of the results based on comparison with source-type solutions occupies Part I and concerns the PME, the HE and the supercritical range of the FDE, $m > m_c$ with $m_c = (n-2)/n$. A complete description of the L^p–L^q effects is produced, and some natural spaces are introduced: the space $\mathcal{M}$ of finite measures and the Marcinkiewicz M^p or weak L^p spaces. A number of topics of general interest is discussed, as indicated in the introduction to that part.

Part II deals with the critical and subcritical range of the FDE, $m \le m_c$. This study offers a large number of novelties, like extinction, backward effects, and delayed boundedness. These novel situations show the sharp contrast between linear and nonlinear equations that occurs in different instances and affects all the basic questions of the theory.

A complete summary of the results obtained so far is contained in Chapter 10. This chapter also summarizes the evolution of Dirac masses in the different exponent ranges, which very much reflects the variation with m of the diffusive power of the equation.

Once the analysis of the PME/FDE has been completed, we devote Part III to extensions and appendices.

As for the first, Chapter 11 contains the application of the same strategies to obtain estimates for the other equations, the p-Laplacian equation and the doubly nonlinear equation. The aim is illustrative of the scope of the method, hence the study of the extension is shorter and less complete. The treatment is necessarily sketchy.

A series of three appendices contains important technical material that is needed but is not in the main line of the book. We hope they will be useful for the reader. We add a final section containing comments and bibliographical notes and end with a brief comment on extensions.

Preference is given to the Cauchy problem, for a question of definiteness, simplicity, and space. This restriction allows us to build a rather complete theory. However, some hints about Dirichlet, Neumann, or local solutions are reflected here and there.

We refer to the monograph [V06] for further details on these problems and also on the peculiarities of solutions with changing sign.

We include at the end of each chapter some additional information on the contents of the different sections, main references, and possible developments, as an orientation for the reader. In doing that we have tried to be concise and fair, and not stray far away from the main subject, but we allow ourselves diversions that might be appealing for the expert or the curious reader, and, of course, we follow our own taste in the matter. We ask for apologies if unintended omissions of important topics arise. Most chapters contain a list of exercises.

Note In obtaining our estimates we will restrict our attention to non-negative solutions and data. Signed solutions can be considered but then the equation must be written as $u_t = \Delta(|u|^{m-1}u)$, or even better $u_t = \nabla \cdot (|u|^{m-1}\nabla u)$ after scaling out the constant m. Restriction to non-negative data is done mostly for convenience, since on the one hand many results will be directly applicable to signed solutions once the non-negative case is settled by using the maximum principle; on the other hand, the physical applications deal in general with the situation $u \geq 0$.

Reading and using the book

The book is aimed at providing an introduction to the subject of smoothing estimates for nonlinear diffusion equations, centred in the study of some model equations and on the Cauchy problem, which is mathematically the most natural in that respect. At the same time, it gives a rather comprehensive account of the state of the art in the case of the PME/FDE, specially the latter one whose theory is still being developed. Consequently, it should be useful as an advanced graduate textbook and also as a source of information for graduate students and researchers. In any case, we have to point out that nonlinear diffusion is a vast field, so that having a text of a manageable size implies a strict selection of material. Many interesting topics and connections have been necessarily omitted for that reason if they played no essential role in the picture we wanted to develop. Some of them are briefly touched on in the sections entitled Comments and historical notes.

As a subject for a PhD course, there are a number of selections depending on purpose and taste. One such selection centres around the supercritical areas contained in Chapters 2 and 3, and some subcritical topics of Chapters 5 *et seq*, plus the inclusion of the p-Laplacian extensions. This option can also be complemented with estimates obtained by an energy approach or other methods as often done in [V06].

Another option is to go as quickly as possible to Chapter 5 and then concentrate on the mysteries of subcritical fast diffusion. There, extensions for further study come from the cited literature.

A basic selection of the text can be combined with advanced topics in related areas. An example is the basic PME theory developed in [V06]. It can be complemented with cases of the dynamical systems approach of [GV03]. Free-boundary problems, kinetic equations and geometrical problems are other options.

As prerequisites, the author has in mind PhD students having followed courses in classical analysis, functional analysis, ODEs and PDEs. Knowing some physics of continuous media or studying the subject in parallel is useful and recommended, but not required. Though the basic part of the text is self-contained, material from other sources is used in the advanced sections that are meant to introduce current problems; in those cases, the pertinent references are carefully indicated and will hopefully serve to introduce the reader to the specialized literature.

Part I
Estimates for the PME/FDE

The first chapter provides necessary preliminaries on functional analysis, comparison results, and the fundamentals of the PME. Additional information is supplied in the appendices at the end.

Chapter 2 discusses the smoothing and decay effects for the porous medium equation, using as a model case the famous Barenblatt solutions that have explicit formulas. This material has been the foundation for all later developments. Important issues are introduced and discussed at length, like scaling techniques, optimal decay, best constants, and the distinction between strong and weak smoothing effects.

Chapter 3 covers the smoothing effects that arise from comparison in Marcinkiewicz spaces, our second major topic. Weak L^p–L^q effects are discussed in Section 3.6.

The short Chapter 4 contains lower estimates (positivity estimates) and Harnack inequalities. It also deals with contractivity, error estimates, and continuous dependence.

1
Preliminaries

The most useful preliminary information is organized in three sections.The first contains the basic facts that we need on concentration and symmetrization, plus a discussion of worst-case strategies and Marcinkiewicz spaces that will play a prominent role in the analysis.

The aim of Parts I and II will be the analysis of smoothing estimates and the related topics that arise for the porous medium equation and its relative, the fast diffusion equation. We devote Section 1.2 to the basic information on that equation that will serve as background material.

Next come the needed comparison results. These results have been discussed in more detail in [Va04b]; the main facts are recalled in Section 1.3 for the reader's convenience since they are essential in the derivation of the estimates.

There is still a technique that plays a big role in the text, namely, the existence of certain types of self-similar solutions which serve as comparison functions. Such analysis will be introduced in Section 3.2 and later on as the need arises.

For the sake of possible extensions, we formulate the preliminary material in terms of signed functions.

1.1 Functional preliminaries

Let Ω a domain in $\mathbf{R}^n$, not necessarily bounded. We denote by $|\Omega|$ the Lebesgue measure of Ω and by $\mathcal{L}(\Omega)$ the set of (classes of) Lebesgue measurable real functions defined in Ω up to a.e. equivalence.

For every function $f \in \mathcal{L}(\Omega)$ we define the distribution function μ_f of f by the formula

$$\mu_f(k) = \text{meas}\{x : |f(x)| > k\}, \tag{1.1}$$

where meas(E) denotes the Lebesgue measure of a set $E \subset \mathbf{R}^n$. We denote by $\mathcal{L}_0(\Omega)$ the subspace of measurable functions in Ω such that $\mu_f(k)$ is finite for every $k > 0$. If Ω has finite measure then $\mathcal{L}_0(\Omega) = \mathcal{L}(\Omega)$, otherwise $\mathcal{L}_0(\Omega)$ contains the measurable functions that tend to zero at infinity in a weak sense. All $L^p(\Omega)$ spaces with $1 \le p < \infty$ are contained in $\mathcal{L}_0(\Omega)$.

1.1.1 Rearrangement

A measurable function f defined in $\mathbf{R}^n$ is called radially symmetric (or radial for short) if $f(x) = \tilde{f}(r)$, $r = |x|$. It is called *rearranged* if it is non-negative, radially symmetric, and $\tilde{f}$ is a non-increasing function of $r > 0$. For definiteness, we also impose that $\tilde{f}$ be left-continuous at every jump point. We will often write $f(x) = f(r)$ by abuse of notation.

A similar definition applies to functions defined in a ball $B = B_R(0) = \{x \in \mathbf{R}^n : |x| < R\}$.

1.1.2 Schwarz symmetrization

For every bounded domain Ω, the *symmetrized domain* is the ball $\Omega^* = B_R(0)$ having the same volume as Ω, i.e.,

$$|\Omega| := \text{meas}\,(\Omega) = \omega_n R^n. \tag{1.2}$$

The precise value ω_n of the volume of the unit ball in $\mathbf{R}^n$ appears in Appendix AI.1. We put $(\mathbf{R}^n)^* = \mathbf{R}^n$. For a function $f \in \mathcal{L}_0(\Omega)$ we define the *spherical rearrangement* of f (also called the symmetrized function of f) as the unique rearranged function f^* defined in Ω^* which has the same distribution function as f, i.e., for every $k > 0$

$$\mu_f(k) := \text{meas}\,\{x \in \Omega : |f(x)| > k\} = \text{meas}\,\{x \in \Omega^* : f^*(x) > k\} \tag{1.3}$$

(we omit the subindex f when it can be inferred from the context). This means that

$$f^*(x) = \inf\,\{k \geq 0 : \text{meas}\{y : |f(y)| > k\} < \omega_n |x|^n\}. \tag{1.4}$$

A rearranged function coincides with its spherical rearrangement. Sometimes the name symmetric decreasing rearrangement is used. For more details, cf. e.g., [B80, Ta76b, Ta76].

The following Hardy–Littlewood formula is well-known and illustrates the relation between f and f^*:

$$\int_{B_R(0)} f^*\,dx = \sup\left\{\int_E |f|\,dx \;:\; E \subset \Omega,\ \text{meas}\,(E) \leq \text{meas}\,(B_R)\right\}. \tag{1.5}$$

There is also an immediate relation between distribution functions and L^p integrals given by the formula

$$\int_\Omega |f|^p\,dx = -\int_0^\infty k^p\,d\mu_f(k) = p\int_0^\infty k^{p-1}\mu_f(k)\,dk. \tag{1.6}$$

Since the distribution functions of f and f^* are identical, equality of integrals

$$\int_\Omega |f|^p\,dx = \int_{\Omega^*} (f^*)^p\,dx \tag{1.7}$$

holds for every $p \in [1, \infty)$.

1.1.3 Mass concentration

The following notion, a variant of the comparison introduced by Hardy and Littlewood, was introduced and used in [Va82] as a basic tool. For every two radially symmetric functions $f, g \in L^1_{\mathrm{loc}}(\mathbf{R}^n)$ we say that f is *more concentrated* than g, $f \succ g$, if for every $R > 0$,

$$\int_{B_R(0)} f(x)\,dx \geq \int_{B_R(0)} g(x)\,dx, \tag{1.8}$$

i.e.,

$$\int_0^R f(r)r^{n-1}\,dr \geq \int_0^R g(r)r^{n-1}\,dr. \tag{1.9}$$

The partial order relationship $\succ$ is called *comparison of mass concentrations*. We can also write $f \succ g$ in the form $g \prec f$, to mean that g is less concentrated than f. A similar definition applies to radially symmetric and locally integrable functions defined in a ball $B = B_R(0)$, and even for radially symmetric Radon measures. When the functions under consideration are rearranged, this comparison coincides with the one introduced by Hardy and Littlewood [BS88], but it does not in general since we do not rearrange the functions prior to comparison. See [Va04b] for a more detailed discussion of the issue.

In any case, for rearranged functions, the comparison of concentrations can be formulated in an equivalent way.

Lemma 1.1 *Let $f, g \in L^1(\Omega)$ be rearranged functions defined in $\Omega = B_R(0)$ for some $R > 0$ and let $g \to 0$ as $r \to R$. Then, $f \succ g$ if and only if the following property holds: for every $k > 0$*

$$\int_\Omega (f(x) - k)_+\,dx \geq \int_\Omega (g(x) - k)_+\,dx\,, \tag{1.10}$$

and if and only if for every convex non-decreasing function $\Phi : [0, \infty) \to [0, \infty)$ with $\Phi(0) = 0$ we have

$$\int_\Omega \Phi(f(t))\,dt \geq \int_\Omega \Phi(g(t))\,dt. \tag{1.11}$$

The result is true if $\Omega = \mathbf{R}^n$, under the assumptions that $f, g \in L^1_{\mathrm{loc}}(\mathbf{R}^n)$ and that $f, g \to 0$ as $r \to \infty$.

We refer to [Va04b] for more details about the preceding topics.

1.1.4 Worst-case strategy. Measures and Marcinkiewicz spaces

The derivation of practical a priori estimates in later sections will be much simplified by using a strategy based on solving more or less explicitly the worst case of a family of problems. This is based on the remark that the relation $\succ$ admits maximal elements when restricted to convenient subclasses of $\mathcal{L}_0(\mathbf{R}^n)$.

Thus, if we consider the subclass of (radially symmetric) and non-negative functions in $L^1(\mathbf{R}^n)$ with fixed mass, i.e., $\int f(x)dx = M$, then we find that a most concentrated element exists, though it lies outside the original functional class. Indeed, it is given by the Dirac mass $M\delta(x)$, which belongs to the space $\mathcal{M}(\mathbf{R}^n)$ of bounded non-negative Radon measures. Hence, this latter space is a more natural candidate when dealing with such problems. Fortunately, the existence and uniqueness theory for the filtration equation has been extended to accept data in this class; but we point out that this convenient extension is not strictly necessary, since comparison arguments between solutions with data in $L^1(\mathbf{R})$ and the worst-case solution with data a Dirac mass can be justified by approximation as long as the latter solution exists.

In the same way, when we consider a subclass of functions in $L^p(\mathbf{R}^n)$ for some $1 < p < \infty$ with a fixed bound for the norm, we cannot find a most concentrated element. However, this task is easy when we extend the class into the corresponding Marcinkiewicz space $M^p(\mathbf{R}^n)$ which is defined as the set of functions $f \in L^1_{\rm loc}(\mathbf{R}^n)$ such that

$$\int_K |f(x)|dx \le C|K|^{(p-1)/p}, \tag{1.12}$$

for all subsets K of finite measure, cf. [BBC75]. The minimal C in (1.12) gives a norm in this space, i.e.,

$$\|f\|_{M^p(\mathbf{R}^n)} = \sup\left\{\text{meas}\,(K)^{\frac{1}{p}-1}\int_K |f|\,dx\,:\,K\subset\mathbf{R}^n,\ \text{meas}\,(K)>0\right\}. \tag{1.13}$$

Since functions in $L^p(\mathbf{R}^n)$ satisfy inequality (1.12) with $C = \|f\|_{L^p}$ (by Hölder's inequality), we conclude that $L^p(\mathbf{R}^n) \subset M^p(\mathbf{R}^n)$ and $\|f\|_{M^p} \le \|f\|_{L^p}$. The Marcinkiewicz space is a particular case of Lorentz space, precisely $L^{p,\infty}(\mathbf{R}^n)$, and is also called weak L^p space. An equivalent norm is described in Appendix AI.2.

We remark that $L^p(\mathbf{R}^n)$ and $M^p(\mathbf{R}^n)$ are different spaces. Indeed, the function given by

$$U_p(x) = A\,|x|^{-n/p} \tag{1.14}$$

is the most typical representative of $M^p(\mathbf{R}^n)$ and is not in $L^p(\mathbf{R}^n)$. Its norm is

$$\|U_p\|_{M^p} = A\,\kappa_p, \qquad \kappa_p = \frac{p\,\omega_n^{1/p}}{p-1}. \tag{1.15}$$

Marcinkiewicz spaces will be of much use for us because of their good combination with the property of concentration. Indeed, it is easy to see that U_p is the most concentrated element of $M^p(\mathbf{R}^n)$ having a given norm. Indeed,

$$\int_{B_R} U_p(x)\,dx = \|U_p\|_{M^p}|B_R|^{(p-1)/p}$$

for all balls B_R, $R > 0$, and the equality turns into $\le$ when we replace B by a set E of finite measure in view of the symmetrization inequality. As regards the comparison of concentrations, the following result holds:

Lemma 1.2 *Function U_p is more concentrated that any function $f \in L^p(\mathbf{R}^n)$ having L^p norm equal or less than $\|U_p\|_{M^p(\mathbf{R}^n)}$.*

Marcinkiewicz spaces will appear prominently in the extinction analysis, cf. Chapter 2 and in the boundedness results of Chapter 6. Let us introduce an interesting functional that measures how much the Marcinkiewicz space differs from the Lebesgue space, and will be used in an essential way in Chapter 6. It is as follows: for an M^p-function f we define

$$\mathrm{N}_p(f) = \lim_{A\to\infty} \|(|f| - A)_+\|_{M^p}. \tag{1.16}$$

This limit exists since the family $(|f| - A)_+$ is non-increasing as $A \to \infty$. Actually, the definition only needs f to be such that $(|f| - A)_+ \in M^p(\mathbf{R}^n)$ for some $A > 0$, as happens with functions $f \in M^p(\mathbf{R}^n) + L^\infty(\mathbf{R}^n)$. Note that the limit is not zero in the case of example (1.14); actually, the limit is still $\mathrm{N}_p(U_p) = A\,\kappa_p$. It is also clear that $\mathrm{N}_p(f)$ is zero for $f \in L^p(\mathbf{R}^n)$. This remark shows that $L^p(\mathbf{R}^n)$ is not dense in $M^p(\mathbf{R}^n)$ for $1 < p < \infty$.

We discuss in more detail the theory of these spaces in Appendix AI.2. The above definitions apply when the functional spaces are defined over an open set $\Omega \subset \mathbf{R}^n$, but we will not be using such extension in this work.

1.1.5 Comparison of maximal monotone graphs (diffusivities)

The filtration equation (2) in the Introduction contains a nonlinearity φ, that is supposed to be a non-decreasing continuous real function. More generally, it can be a maximal monotone graph. We recall in Appendix AI.4 this concept for the reader's convenience. We will be interested in comparing the concentrations of solutions of two equations with different φ. We have introduced in [Va04b] the following concept.

Definition 1.1 We say that a maximal monotone graph φ_1 *is weaker than another one* φ_2, and we write $\varphi_1 \prec \varphi_2$, if they have the same domains, $D(\varphi_1) = D(\varphi_2)$, and there is a contraction $\gamma : \mathbf{R} \to \mathbf{R}$ such that

$$\varphi_1 = \gamma \circ \varphi_2 . \tag{1.17}$$

By contraction we mean $|\gamma(a) - \gamma(b)| \le |a - b|$. This implies in particular φ_1 must have horizontal points (or horizontal intervals) at the same values of the argument as φ_2, and maybe some more. We also assume that φ_1 does not accept vertical intervals (i.e., it is one-valued). Note that for smooth graphs condition (1.17) just means that

$$\varphi_1'(s) \le \varphi_2'(s), \qquad \text{for every } s \in D(\varphi_2), \tag{1.18}$$

which is easier to remember or to manipulate. In parabolic problems, $\varphi'(u)$ is interpreted as the diffusivity, so that relation (1.18) can be phrased as: *φ_1 is less diffusive than φ_2.* This explains why it will be important in our evolution analysis.

1.2 Preliminaries on the PME and the FDE

The following chapters contain a detailed study of the *time decay* and *smoothing estimates* that can be obtained for the PME/FDE using the machinery introduced so far and the strategy of the worst case. This is a topic in which the proposed method proves to be quite effective. The exponent m is allowed to cover the range $m \in \mathbf{R}$. In order to keep the text within some reasonable bounds, we assume that there is no forcing term, $f = 0$.

1.2.1 Basic properties of the porous medium and fast diffusion flow

Before proceeding with estimates and proofs, let us recall some basic facts that we will be using about the theory of the PME/FDE. The basic theory of the PME in its complete form $u_t = \Delta(|u|^{m-1}u) + f$ is contained in the monograph [V06] to which we refer for further details; it covers not only the Cauchy problem for non-negative solutions but also the Dirichlet and Neumann boundary value problems for signed solutions. We recall next the main facts that we need for the plain case $u_t = \Delta u^m$ that we will be using here.

(i) WELL-POSEDNESS. Assume that $m > 0$ and consider the question of solving $u_t = \Delta u^m$. The method of implicit discretization in time allows us to construct for every data $u_0 \in L^1(\mathbf{R}^n)$ a so-called mild solution $u(t) = S_t u_0 \in C([0, \infty) : L^1(\mathbf{R}^n))$. Actually, the maps $u_0 \mapsto u(t) = S_t u_0$ form a semigroup of contractions in the space $X = L^1(\mathbf{R}^n)$. Moreover, for the equation with $m > 0$ the concepts of mild, weak, and strong solution are shown to be equivalent, and such a solution is unique, continuous, and depends continuously on the data, cf. [BC81, BCP]. We have also pointed out that the equation must be written as $u_t = \Delta(|u|^{m-1}u)$ if negative values are also involved. As already mentioned, we will restrict our attention to non-negative solutions and data. Estimates for signed solutions will follow immediately by using the maximum principle.

We will be mostly working with solutions with initial data in the Lebesgue spaces $L^p(\mathbf{R}^n)$, $1 \le p \le \infty$, and the Marcinkiewicz spaces $M^p(\mathbf{R}^n)$, $1 < p < \infty$. One option open to us is to formulate our estimates for solutions having initial data in the intersection of these spaces with $L^1(\mathbf{R}^n)$, thus dispensing with the more general existence theory, which is not needed for our presentation in a strict sense. Actually, our estimates can be first derived under such restrictions and then used to build a theory in extended functional spaces. However, since a general existence theory is now well established, it seems to us convenient to make use of it in formulating our results, since it allows us to show their full scope.

Indeed, the concept of solution can be extended to wider classes of initial data. Always keeping the restriction to non-negative data and solutions, if the extension is performed inside the large space $\mathcal{Z} = L^1_{\rm loc}(\mathbf{R}^n)^+$, then the optimal class for the PME is known to be the set of data

$$\mathcal{X}_m = \left\{ u_0 \in \mathcal{Z} : \int_{B_R(0)} u_0(x)\, dx = O(R^{p_m}) \right\}, \quad p_m = n + \frac{2}{m-1}, \tag{1.19}$$

where we use Landau's big Oh notation in the limit $R \to \infty$; it must be replaced by small o if the solution is to exist globally in time.

The analogous restriction for the heat equation is the well-known condition on square exponential growth as $R \to \infty$, while no condition is needed if $m \in (0, 1)$, and then $\mathcal{X}_m = \mathcal{Z}$. Actually, existence and uniqueness of a weak solution is true for initial data in that space, cf. [HP85]. Moreover, the weak solution depends continuously with respect to the topology of weak convergence in $L^1_{\text{loc}}(\mathbf{R}^n)$. On the practical side, the above setting includes the non-negative cones of the Lebesgue spaces $L^p(\mathbf{R}^n)$ and Marcinkiewicz spaces $M^p(\mathbf{R}^n)$ that we will be using.

From the point of view of regularity, the best situation corresponds to the exponent range where m is above the critical value, $m_c < m < 1$ with

$$m_c = (n-2)/n, \tag{1.20}$$

or $0 < m < 1$ if $n = 1$. Then, it is proved in [HP85] that weak solutions with data in $L^1_{\text{loc}}(\mathbf{R}^n)$ become immediately locally bounded (local regularizing effect), and consequently positive and C^∞ smooth in $Q = \mathbf{R}^n \times (0, \infty)$, thus ensuring that the weak theory is locally a classical theory in that range.

Such regularizing effect fails for exponents below the critical value as we will show in detail in our study. We note that our weak solutions with data in $\mathcal{Z} = L^1_{\text{loc}}(\mathbf{R}^n)$ can be obtained as monotone limits of smooth solutions with bounded initial data, but they need not be bounded.

(ii) MEASURES AS DATA. The existence theory can be extended to the wider space $\mathcal{Z}_1$ of locally finite Radon measures and then the initial-value problems are well-posed if the same type of growth condition is used if $m > 1$ or $m = 1$. For exponents $m < 1$ no growth condition is needed if m is above the critical value m_c. The constructed solutions in those ranges are integrable functions (not only measures) for all $t > 0$.

As a further extension, unique generalized solutions with general (non-negative) Borel measures have been constructed in the good range $m_c < m < 1$ by Chasseigne-Vázquez [ChV02]. We recall that Borel measures are not locally finite. The corresponding solutions are not bounded at the places where the initial measure is not locally finite (i.e., they exhibit permanent singularities).

On the other hand, it was discovered by Brezis and Friedman [BF83] that Dirac deltas are not diffused by the FDE with exponent equal to or below the critical value m_c, which marks an important division line in the theory that we will develop. The theory of the FDE for exponents $m \le m_c$ will be discussed as part of the study in Chapter 5 *et seq.*

(iii) COMPACTNESS. It is also known that families of solutions of the PME or FDE that are non-negative and uniformly bounded are uniformly Hölder continuous with certain Hölder exponent that depends only on m and n (and is not always known). The precise result says that there exist an exponent $\mu \in (0, 1)$ and a constant $C_h = C_h > 0$, both depending only on m and n, such that any solution of the PME $\tilde{u}(x, t)$ defined in the strip $\{(x, t) : 1/2 \le t \le 1\}$ and bounded in it by 1 satisfies

$$|\tilde{u}(x, 1) - \tilde{u}(x', 1)| \le C_h |x - x'|^\mu. \tag{1.21}$$

Compactness in other functional classes depends on the smoothing effects that we can prove. Actually, a smoothing effect can be viewed as a first step in the way to regularity and compactness. It must be said that, in a logical sense, all these regularity results come after we have proved that solutions are bounded, which is the main concern of the present text.

(iv) SCALING. The PME and the FDE share with the HE a powerful property inherited from the power-like form of the nonlinearity. This is the invariance under a transformation group of homotheties, usually known *the scaling group*. Indeed, whenever $u(x,t)$ is a (weak, classical, mild) solution of the equation, the rescaled function

$$\tilde{u}(x,t) = K\, u(Lx, Tt) \tag{1.22}$$

is also a solution if the three real parameters K, L, T are tied by the relation

$$K^{m-1}L^2 = T. \tag{1.23}$$

We get in this way a two-parameter family of transformed solutions. We can further restrict the family to a one-parameter family by imposing another condition; a typical example is the condition of preserving the L^p norm of the data or the solution. In the case of the data, it reads

$$\int_{\mathbf{R}^n} u^p(x,0)\,dx = \int_{\mathbf{R}^n} K^p\, u^p(Lx,0)\,dx,$$

which implies the condition $K^p = L^n$. This allows us to determine two parameters in terms of the third, but for some exceptional cases. Remarkably, such cases play a special role in the theory and will receive due attention.

We will use scaling and special solutions that are scaling-invariant, in combination with symmetrization and mass comparison, as basic tools in obtaining the estimates of the following chapters. This is in line with the fact that scaling arguments are well-known and very successful in the applied literature. For a detailed study of self-similarity and scaling in mathematics and mechanics we refer to the classical books by G. Barenblatt [Ba87, Ba79]. The ideas are better known in theoretical physics as the renormalization group.

(v) CONTRACTION IN L^1 AND DECAY OF L^p NORMS. If the data belong to an L^p space then the solution belongs to the same space for all $t > 0$ and we have

$$\|u(t)\|_p \le \|u_0\|_p. \tag{1.24}$$

This is valid for all $p \in [1, \infty]$ and even for Orlicz norms.

On the other hand, we have a stronger contraction property. Given two solutions u_1 and u_2 in $C([0,T]; L^1(\mathbf{R}^n))$ and times $0 \le t_1 \le t_2 \le T$, we have

$$\|u_1(t_2) - u_2(t_2)\|_1 \le \|u_1(t_1) - u_2(t_1)\|_1. \tag{1.25}$$

We point out that the formula applies whenever the right-hand side is finite and u_{01}, u_{02} belong to the existence class, even if they do not belong to $L^1(\mathbf{R}^n)$. Note also that this contraction property is not true for any $p > 1$ if $m \neq 1$.[1]

[1] In this text, the word contraction is meant as non-strict contraction, i.e., using the sign $\le$ in formula (1.25). We provide in [V06] a proof of the lack of contraction in some L^p norms, $p > 1$.

(vi) UNIVERSAL ESTIMATES. The scaling property of the equation is responsible for the existence and simple form of some a priori *universal estimates*, by which we mean that they are valid for all non-negative solutions of the equation (at least solutions defined by standard weak theories and their limits). The first of them is due to Aronson and Bénilan [AB79] and reads

$$\Delta u^{m-1} \geq -\frac{C_1}{t} \tag{1.26}$$

for all $m > 1$. The sharp constant is known, $C_1 = n(m-1)/m(n(m-1)+2)$. The estimate extends to $m = 1$ in the form $\Delta \log u \geq -n/2t$; it also holds in the FD range $1 > m > (n-2)/n$, but the inequality sign is reversed.

Another universal estimate is due to Bénilan and Crandall [BC81b] and applies to all non-negative weak solutions and their limits even if $m < 1$. It reads

$$(m-1)u_t + u \geq 0. \tag{1.27}$$

These are famous pointwise inequalities on which much of the theory is based, cf. [V06], but we will not have a strong use for them.

On the concept of fast and slow diffusion Let us make a warning about terminology. The equation for $m < 1$ is called the fast diffusion equation, while the equation for $m > 1$ is often referred to as *slow diffusion*. Since the diffusivity coefficient is $D(u) = u^{m-1}$, the label is appropriate when $0 < u < 1$ (or for $0 < |u| < 1$ in the signed equation). However, for larger values of u the situation is rather the converse : $m > 1$ is actually fast diffusion and $m < 1$ is slow diffusion. This confusing situation is a typical problem created by power functions. It will be advisable for the reader to bear it in mind when interpreting some of the results, like comparison of decay rates.

1.2.2 Range $m \leq 0$. Superfast diffusion. Modified equation

So far, we have assumed $m > 0$, and this condition will be kept in the next chapters. However, it will be convenient to say some words about the case $m \leq 0$ which plays an important role in this text. Since the diffusivity is given by $D(u) = u^{m-1}$, the range $m \leq 0$ is sometimes called *superfast diffusion*, a term that is justified when applied to small densities $u \sim 0$. It is also called very singular diffusion.

There is a theory of existence and uniqueness for the FDE in that range of exponents on the condition that we write the equation in the slightly modified form

$$u_t = \text{div}\,(u^{m-1}\nabla u) + f, \tag{1.28}$$

in order to keep the parabolicity. The reader will easily check that the original equation is backwards parabolic for $m < 0$, thus ill-posed in principle; note also that for $m > 0$ the modified equation (1.28) obtains after a simple time scaling from equation $u_\tau = \Delta u^m$, the relation between times being $\tau = t/m$. We will use the modified version of the equation in the sequel. The change is inessential when $m > 0$ as for the qualitative results, but it does affect the value of the best constants in the smoothing estimates by a factor m that goes with the time variable; they take anyway a simpler form in the modified version that we will follow in the sequel. The change is essential for $m \le 0$. For $m = 0$ the original equation does not amount to any evolution, while the modified one represents the Ricci flow for conformal surfaces, a quite popular subject in differential geometry nowadays. We will treat the topic in Chapter 8.

Though keeping many of the standard properties common to the range $m > 0$, the theory has a number of peculiar features in the range $m < 0$. A very striking aspect is formed by the non-existence results, like the one established in [Va92], which for $n \ge 3$ says:

If $m \le 0$ there is no solution of the FDE with integrable initial data $u_0 \in L^1(\mathbf{R}^n)$. Similar results hold in dimensions $n = 1$ and 2. Precisely, our a priori estimates are expected to shed light on these phenomena. Anyway, these issues will have to wait until Part II.

1.3 Main comparison results

We have discussed in [Va04b] and [V06] the generation of solutions of evolution equations of the type of the filtration equation by means of implicit discretization in time, making use of the convergence theorems of nonlinear semigroup theory. Here we recall the main facts and the comparison result we will use. We consider the equation,

$$u_t = \Delta\varphi(u) + f, \tag{1.29}$$

where φ is an increasing real function, or more generally a maximal monotone graph (m.m.g.) with $0 \in \varphi(0)$, so that the equation can be degenerate parabolic (if $\varphi'(u) = 0$ somewhere) or singular parabolic (if $\varphi'(u) = \infty$ somewhere).

A theory of existence and uniqueness of solutions of the Cauchy problem for (1.29) with initial data

$$u(x, 0) = u_0(x) \in L^1(\mathbf{R}^n) \tag{1.30}$$

and right-hand side

$$f \in L^1(Q_T), \quad Q_T = (0, T) \times \mathbf{R}^n \tag{1.31}$$

was developed in the early 1970s in the framework of nonlinear semigroups, in the works of Bénilan, Crandall and others [Be72, BCP, CL71, Ev78], and in fact it was one of the motivations in the development of that theory. Briefly stated, the solution of the evolution problem is reduced by means of implicit time discretization to the iterative solution of an associated elliptic problem. We thus obtain an approximate solution defined continuously in space but discretely in time. The scheme works perfectly for the class of nonlinear evolution problems that can be written in the form $u_t + A(u) = f$ and A is a possibly nonlinear and possibly multivalued m-accretive operator, as is our case. Moreover, the famed Crandall–Liggett theorem allows us to pass to the limit in the approximate solutions as the time-step goes to zero to obtain a so-called *mild solution* of the evolution problem.

Starting with paper [Va82], we have developed the theory of symmetrization for the semilinear elliptic equations that appear in the discretization steps; on the other hand, we have shown how to use it in combination with the Crandall–Liggett theorem to derive the following comparison results for the nonlinear parabolic equation (1.29), which are basic in what follows.

Theorem 1.3 *Let u be the mild solution of problem* (1.29)–(1.30) *with data u_0, nonlinearity φ and second member f under the above assumptions. Let v the solution of a similar problem with radially symmetric data $v_0(r) \geq 0$, nonlinearity ψ and second member $g(r,t) \geq 0$. Assume moreover that*

(i) $u_0^* \prec v_0$,
(ii) $\psi \prec \varphi$ and $\varphi(0) = \psi(0) = 0$,
(iii) $f^*(\cdot, t) \prec g(\cdot, t)$ for every $t \geq 0$.

Then, for every $t \geq 0$

$$u^*(\cdot, t) \prec v(\cdot, t). \tag{1.32}$$

In particular, for every $p \in [1, \infty]$ we have comparison of L^p norms,

$$\|u(\cdot, t)\|_p \leq \|v(\cdot, t)\|_p \,. \tag{1.33}$$

The mild solution is the unique solution obtained by the Crandall–Liggett implicit discretization scheme, as explained in detail in [Va04b]. Note that the norms of (1.33) can also be infinite for some or all values of p. Note also the order reversal in condition (ii). The result was first proved in [Va82] for the case where the nonlinearities are the same, $\varphi = \psi$.

There is a second related result that explains that solutions are less concentrated than the data and spread out in time (if there is no forcing term). We see this result as a form of a general principle, that we describe as the law of *decreasing concentration of solutions of nonlinear diffusion.*

Theorem 1.4 *Let $f = 0$ and Let $0 \leq u_0$ be a radially symmetric function in $L^1(\mathbf{R}^n)$. If $u(x,t)$ is the mild-solution of problem* (1.29)–(1.30), *then*

$$u(\cdot, t) \prec u(\cdot, s) \prec u_0 \qquad \text{for } t \geq s \geq 0. \tag{1.34}$$

This is part of a more general principle valid for nonlinear diffusion equations, the law of *decreasing concentration* of solutions in time, that is discussed in detail in [Va04b, Proposition 9.2].

Similar results are true for Dirichlet problems posed in bounded domains, but we will not go into that issue in this text.

1.4 Comments and historical notes

Section 1.1 Most of the definitions are standard. The concept of mass concentration was introduced into the study of a priori estimates for parabolic PDEs in [Va82]. We refer to [Va04b] for a full study of the elliptic results on symmetrization and concentration, and for references to the relevant literature in the field.

Section 1.2 The porous medium equation and the fast diffusion equation arise in many applications in physics, chemistry, biology, and engineering. The common idea is that in many diffusion processes the diffusion coefficient depends on the unknown quantities (concentration, density, temperature, etc.) of the diffusion model. The PME owes its name to the modelling of the flow of gases in porous media [Le29, Mu37]. Its use in high-temperature physics is documented in [ZR66]. Earlier on, Boussinesq proposed it in a problem of groundwater infiltration [Bo03], 1903. Filtration through natural rocks is studied for instance in [BER90, Bea72]. The mathematical study of the PME can be said to have begun in earnest with the work of Zeldovich, Kompanyeets and Barenblatt [ZK50, Ba52], on the existence of source-type solutions in 1950–52. The study of well-posedness in the framework of weak solutions was initiated by O. Oleinik and her group around 1958 [OKC58].

General references for the Mathematics of the PME are given at the beginning of the section. Further information on the Mathematics of the PME can be obtained from [V06]. See also the previous surveys [Ar86, Ka87, Pe81, V92b] and the book [WZYL]. A short updated summary of basic results for the PME is contained in the book [GV03].

Applications of the FDE have been proposed in different areas. The equation appears in plasma physics, the Okuda–Dawson law corresponds to $m = 1/2$, [OD73]. King [Ki88] studies the case $0 < m < 1$ in a model of diffusion of impurities in silicon. The equation in dimensions $n \geq 3$ with exponent $m = (n-2)/(n+2)$ has an important application in geometry (the Yamabe flow) that we study in Section 7.5. References for applications of the equation with exponents $m \leq 0$ will be given in the final sections of Chapters 8 and 9.

For the mathematics of the FDE a basic reference is the already quoted Herrero-Pierre [HP85]. Important sources are also the work of Dahlberg–Kenig [DK88] and Pierre [Pi87], who study weak solutions for the FDE with measures as initial data. A main reference for the use of similarity methods in the study of fast diffusions is Kings' [Ki93].

Central to the understanding of these equations is the corresponding Cauchy problem, on which we focus our attention in this text. However, results on boundary-value problems with different types of boundary conditions (Dirichlet, Neumann, mixed, nonlinear, ... homogeneous or not) are abundant now in the literature and reflect aspects of interest in the theory and the applications.

Information about more general nonlinear parabolic equations can be found in Kalashnikov's survey paper [Ka87].

Section 1.3 The theory of nonlinear semigroups was developed by many authors in the 1960s and 1970s. The works of Bénilan, Brezis, Crandall and Evans are important for our context. Unfortunately, the book [BCP], which contains a wealth of results on the subject, has not been published. We discuss at length that issue in Chapter 10 of [V06]. A useful presentation is also given as an appendix in [ACM04].

The comparison results we present here originate from [Va82] and are established in whole detail in [Va04b].

There are many possible applications of the comparison techniques. We summarize the process by saying that we want to obtain estimates valid for large classes of solutions and equations by performing some simple calculations in representative cases. A first step consists in replacing estimates for n-dimensional solutions by estimates for easier one-dimensional problems by reducing ourselves to radially symmetric situations. This is the crux of the symmetrization technique. We further simplify the problem by showing how to compare radial solutions with different initial data u_0, second member f and constitutive nonlinearity φ. In the problems that follow we can find a *worst case* in suitable classes of solutions, and then the estimate consists of just reading the information from that worst-case solution.

2

Smoothing effect and time decay. Data in $L^1(\mathbb{R}^n)$ or $\mathcal{M}(\mathbb{R}^n)$

We are ready to proceed with the systematic derivation of the smoothing estimates. We focus our attention on the study of the *porous medium equation* (PME) without forcing term, that we always write in the modified form

$$u_t = \operatorname{div}(u^{m-1}\nabla u). \tag{2.1}$$

After examining the source-type solutions in great detail in Section 2.1, we prove in Section 2.2 the smoothing effect into L^∞, originally due to Bénilan [Be76] and Véron [Ve79]. We supply a different, direct proof that allows for the explicit calculation of the best constants and is the first of the main results of this text. This chapter covers the original exponent range, $m > 1$. The same method also covers the heat equation $m = 1$, where we obtain well-known results by a technique which is quite different from the usual representation analysis. We even cover the fast diffusion equation, $m < 1$, but only in the range $m > m_c$, where $m_c = (n-2)/n$ is an important critical number; the exponent restriction is an essential limitation (i.e., it is not only technical) that will be overcome in Part II by new methods giving rise to a different kind of results, a totally different functional world. The easy extension to effects into L^p with $1 < p < \infty$ is left to Section 2.7.

The interplay between scaling invariance and decay exponents is carefully explained in Section 2.3; this is a fundamental property of scale-invariant equations. Section 2.4 explains the difference between strong and weak smoothing effects, an often ignored question which plays an important role in the sequel.

Though we concentrate on equations with power nonlinearities, the practice of nonlinear diffusion leads to many filtration models where the nonlinearity is only approximately power-like. A comparison result to deal with this possibility is discussed in Sections 2.5 and 2.6. In Section 2.8 we briefly touch on the question of actual asymptotic behaviour, which deserves a text of its own. Finally, Section 2.9 treats the interesting limit $m \to \infty$, so-called mesa problem.

Sections 2.1–2.4 and 2.8 are recommended for a basic reading.

2.1 The model. Source-type solutions

A main point of our analysis is that comparison with the corresponding worst case allows us to derive the quantitative expression of these phenomena with exact exponents (rates), and also to obtain the best constants in the inequalities (in other words, we solve *optimal problems*).

In the case of the porous medium equation, there is a worst case with respect to the symmetrization-and-concentration comparison theorem of Section 1.3 in the class of solutions with the same initial mass $\|u_0\|_1 = M$. It is just the solution U with initial data a Dirac mass. This solution exists for $m > 1$ and is explicitly given by

$$U(x,t;M) = t^{-\alpha} F(x/t^{\alpha/n}), \quad F(\xi) = (C - k\,\xi^2)_+^{\frac{1}{m-1}} \tag{2.2}$$

where

$$\alpha = \frac{n}{n(m-1)+2}, \qquad k = \frac{(m-1)\alpha}{2n}. \tag{2.3}$$

It is called the source solution because it has a Dirac delta as initial trace,

$$\lim_{t\to 0} u(x,t) = M\,\delta_0(x). \tag{2.4}$$

The remaining parameter $C > 0$ in formula (2.2) is in principle arbitrary; it can be uniquely determined by the mass condition $\int U dx = M$, which gives the following relation between the 'mass' M and C:

$$M = d\,C^\gamma, \quad d = n\,\omega_n \int_0^\infty (1 - k\,y^2)_+^{1/(m-1)}\, y^{n-1}\,dy, \quad \gamma = \frac{n}{2(m-1)\alpha} \tag{2.5}$$

(d and γ are functions of only m and n; the exact calculation of d will be performed later). This well-known solution is usually called the *source solution* or *Barenblatt solution*, since Barenblatt performed a complete study of these solutions in 1952. The name *ZKB solution* often used in this text honours the work of Zel'dovich and Kompanyeets who supplied the first example. Using the mass as parameter, we denote it by $U(x,t;M)$ or $U_m(x,t;M)$. It is useful to write the complete expression as

$$\boxed{U_m(x,t;M)^{m-1} = \frac{(C\,t^{2\alpha/n} - k\,x^2)_+}{t}} \tag{2.6}$$

with $C = a(m,n)M^{2(m-1)\alpha/n}$. We can pass to the limit $m \to 1$ (with a fixed choice of the mass M) and obtain the fundamental solution of the heat equation,

$$\boxed{E(x,t) = M\,(4\pi t)^{-n/2}\,\exp(-x^2/4t)} \tag{2.7}$$

Therefore, $E(x,t;M) = U_1(x,t;M)$. Note the difference: U_m has compact support in the space variable for all $m > 1$, while E is positive everywhere with exponential tails at infinity. See Figure 2.1.

It was realized that the source solution also exists with many similar properties as long as $\alpha > 0$, i.e., it can be extended to the fast diffusion equation, $m < 1$, but only in the range $m_c < m < 1$, with $m_c = 0$ for $n = 1, 2$, $m_c = (n-2)/n$ for $n \geq 3$.

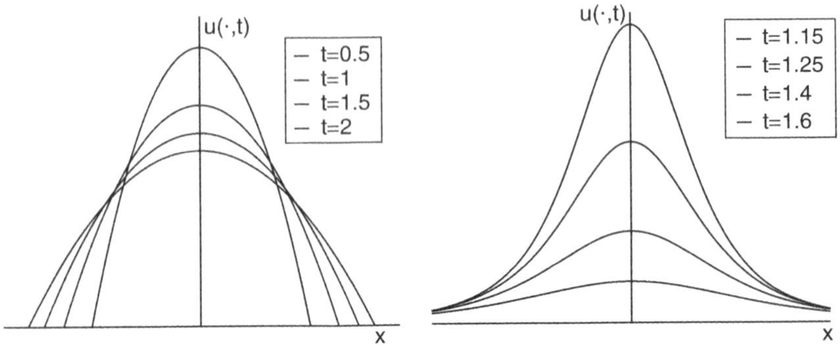

Figure 2.1 Left: evolution of the source solution for $m > 1$ with a free boundary. Right: the FDE source solution for $m < 1$ with a fat tail (polynomial decay). These profiles are the nonlinear alternative to the Gaussian profiles.

Formula (2.2) is basically the same, but now $m - 1$ and k are negative numbers, so that U_m is everywhere positive with a power-like tail at infinity. More precisely,

$$U_m(x, t; M) = t^{-\alpha} F(x/t^{\alpha/n}), \quad F(\xi) = (C + k_1 \xi^2)_+^{-\frac{1}{1-m}}. \tag{2.8}$$

with same value of α and $k_1 = -k = (1 - m)\alpha/2n$. The complete expression is

$$\boxed{U_m(x, t; M)^{1-m} = \frac{t}{C\, t^{2\alpha/n} + k_1\, x^2}, \qquad C = a(m, n) M^{-2(1-m)\alpha/n}} \tag{2.9}$$

The main difference in the ZKB profiles in the different ranges is probably the shape at infinity, which reflects the propagation mode. While for $m > 1$ the ZKB solutions are compactly supported, for $m = 1$ the Gaussian kernel has quadratic exponential decay, and for the fast diffusion range $m_c < m < 1$ we have profiles with an algebraic space decay, $u(x, t) \approx C(t)|x|^{-2/(1-m)}$, which are called in the statistical literature *fat tails*, *robust tails*, and also *overpopulated tails*. The difference between the tails for $m = 1$ and $m < 1$ is not only a question of decay rate. Thus, the tails for $m < 1$ behave in a universal way

$$U_m(x, t) \sim \left(\frac{t}{k_1 x^2}\right)^{\frac{1}{1-m}} \quad \text{as } |x| \to \infty, \tag{2.10}$$

whereas in the heat equation case the behaviour is $E(x, t) \sim M\,(4\pi t)^{-n/2} e^{-x^2/4t}$, an expression that depends on the total mass of the solution.[1] A related observation that will have a consequence in Chapter 5 is the following: in the fast diffusion case we can

[1]But note that this behaviour would be universal in logarithmic variables, $\log E(x, t) \sim -x^2/4t$. In any case, the difference has consequences for the physics and the theory.

put the constant $C = 0$ and still get a non-trivial solution. The 'solution' that we get corresponds to mass $M = \infty$ and reads

$$U_m(x, t; \infty) = \left(\frac{t}{k_1 x^2}\right)^{\frac{1}{1-m}}. \tag{2.11}$$

This singular solution has a standing non-integrable singularity at $x = 0$ and does not decay at all. In fact, it has a positive time derivative everywhere away from the singularity. It does not take part in the description of the asymptotic analysis of weak solutions as $t \to \infty$ though it is exact to describe the behaviour as $|x| \to \infty$ as (2.10) indicates. This solution is called infinite point-source solution, briefly IPSS, in [ChV02], where it is used to describe the behaviour of the class of continuous extended solutions for $m_c < m < 1$.

2.2 Smoothing effect and decay with L^1 functions or measures as data. Best constants

Using the existence and properties of the source solutions and the comparison theorems, we get the following result.

Theorem 2.1 *Let u be the solution of equation* (2.1) *in the range $m > m_c$ with initial datum $u_0 \in L^1(\mathbf{R}^n)$. Then, for every $t > 0$ we have $u(t) \in L^\infty(\mathbf{R}^n)$ and moreover there is a constant $c(m, n) > 0$ such that*

$$|u(x, t)| \le c(m, n)\, \|u_0\|_1^\sigma\, t^{-\alpha}, \tag{2.12}$$

with α given in (2.3) *and $\sigma = 2\alpha/n$. The best constant is attained by the ZKB solution and is given by formulas* (2.13), (2.14), and (2.16) *below.*

The same result holds when u_0 belongs to the space $\mathcal{M}(\mathbf{R}^n)$ of bounded and non-negative Radon measures if $\|u_0\|_1$ is replaced by $\|u_0\|_{\mathcal{M}(\mathbf{R}^n)}$.

Proof (i) It is clear that the worst case with respect to the symmetrization-and-concentration comparison in the class of solutions with the same initial mass M is just the ZKB solution U with initial data a Dirac mass, $u_0(x) = M\delta(x)$. We are thus reduced to perform the computation of the best constant for the ZKB solution. We have

$$\|U(t)\|_\infty = C^{1/(m-1)} t^{-\alpha} = d^{-2\alpha/n} M^{2\alpha/n} t^{-\alpha}.$$

Computing d is an exercise involving Euler's beta and gamma functions, see Appendix AI.1 for the basic formulas. For $m > 1$ we obtain

$$d = k^{-n/2} n\omega_n \int_0^1 (1 - s^2)^{1/(m-1)} s^{n-1} ds = \frac{1}{2} k^{-n/2} n\omega_n B\left(\frac{n}{2}, \frac{m}{m-1}\right),$$

and for $m < 1$

$$d = k^{-n/2} n\omega_n \int_0^\infty (1 + s^2)^{-1/(1-m)} s^{n-1} ds = \frac{1}{2} k^{-n/2} n\omega_n B\left(\frac{n}{2}, \frac{1}{m-1} - \frac{n}{2}\right).$$

We thus conclude that inequality (2.12) – (2.13) holds with the precise constant

$$c(m,n) = \left(\frac{\alpha(m-1)}{2n\pi}\left\{\frac{\Gamma(m/(m-1)+n/2)}{\Gamma(m/(m-1))}\right\}^{2/n}\right)^{\alpha}. \tag{2.13}$$

for $m > 1$, and

$$c(m,n) = \left(\frac{\alpha(1-m)}{2n\pi}\left\{\frac{\Gamma(1/(1-m))}{\Gamma(1/(1-m)-n/2))}\right\}^{2/n}\right)^{\alpha}. \tag{2.14}$$

for $m_c < m < 1$. There are some interesting cases worth commenting. First, the expression is quite simple for $n = 2$, since $\alpha = 1/m$, and an immediate calculation gives

$$c(m,2) = (4\pi)^{-1/m}. \tag{2.15}$$

On the other hand, taking the limit in both expressions (for $m > 1$ and $m < 1$) as $m \to 1$ we get the best constant for the L^1–L^∞ effect for the heat equation

$$c(1,n) = (4\pi)^{-n/2}. \tag{2.16}$$

Finally, $\lim_{m\to\infty} c(m,n) = 1$ for all n, while there is an alternative in the limits in the lower end: $\lim_{m\to m_c} c(m,n) = \infty$ for $n \geq 3$, while the same limit is 0 for $n = 2$. For $n = 1$ see Subsection 2.2.1 below.

(ii) Actually, there is a difficulty in taking U as a worst case in the comparison, namely that $U(x,0;M)$ is not a function but a Dirac mass. We can solve this technical problem by extending the theory to bounded measures as initial data, which seems the most natural way, and indeed such a theory has been developed and offers no problem in the present context, cf. [Pi82]. However, we prefer to stay at a more elementary level and overcome the difficulty by approximation.

We take first a solution with bounded initial data, $u_0 \in L^1(\mathbf{R}^n) \cap L^\infty(\mathbf{R}^n)$. We then replace $U(x,t;M)$ by a slightly delayed function $U(x,t+\tau;M)$, which is a solution with initial data $U(x,\tau;M)$, bounded but converging to $M\delta(x)$ as $\tau \to 0$. It is then clear that for a small $\tau > 0$ such solution is more concentrated than u_0. From the comparison theorem we get

$$|u(x,t)| \leq \|U(\cdot,t+\tau;M)\|_\infty = c(m,n)\, M^\sigma\, (t+\tau)^{-\alpha}, \tag{2.17}$$

which of course implies (2.12). The result for general L^1 data or measures follows by approximation and density once it is proved for bounded L^1 functions. □

Correction for the other version of the PME If the PME is written in the 'usual form' $u_t = \Delta u^m$, the only difference with the previous calculation is the constant m that multiplies the right-hand side. There is a simple way to take into account this constant, which consists in incorporating it into the time variable and writing the usual equation in the form

$$\frac{\partial u}{\partial t'} = \nabla \cdot (u^{m-1}\nabla u), \quad t' = mt. \tag{2.18}$$

The above formulas apply exactly to this formulation with t replaced by t'. In this way, the ZKB is corrected in the value of k that becomes

$$k = \frac{(m-1)\alpha}{2nm}, \tag{2.19}$$

and Theorem 2.1 holds with $c(m,n)$ of formula (2.12) replaced by

$$c'(m,n) = c(m,n)/m^{\alpha}. \tag{2.20}$$

2.2.1 *Singular case in one dimension*

The critical exponent of the fast diffusion equation in dimension $n = 1$ is formally $m_c = (n-2)/n = -1$, and not zero (the lower limit of the m-range that we have assumed in principle). The extra range $-1 < m \leq 0$ has singular diffusivity and exhibits peculiar properties that will be reviewed in Chapter 8; one of them is that it does not have uniqueness of classical solutions for the Cauchy problem. This is a very striking property, since we are dealing with positive, finite-mass solutions of a simple diffusion process. The range has been studied in [ERV88], where existence and properties of the class of maximal solutions are discussed. A maximal solution is characterized by the property of conservation of mass in time.

Now, for $n = 1$ we can check that the ZKB solutions exist for all $m > -1$ for the modified equation (2.1). i.e., we can extend the range to $-1 < m \leq 0$. These solutions read

$$(U_m(x,t))^{1-m} = \frac{t}{a_m\,(t/M^{1-m})^{2/(1+m)} + k|x|^2}, \qquad k = \frac{1-m}{2(1+m)}. \tag{2.21}$$

They are maximal solutions, hence, they may be used as upper bounds in the comparison theorems of the symmetrization-concentration method. The smoothing effect L^1–L^∞ is proved for all solutions of this equation with the same proofs, and same formulas. Thus, Theorem 2.1 is true with $\alpha = 1/(m+1)$ and $\sigma = 2/(m+1)$, and formula (2.14) becomes

$$c(m,1) = \left(\frac{(1-m)}{2\pi(m+1)}\left\{\frac{\Gamma(1/(1-m))}{\Gamma(1/(1-m)-1/2))}\right\}^2\right)^{1/(m+1)}. \tag{2.22}$$

Note that for the special logarithmic case, $m = 0$, $u_t = (\log u)_{xx}$, which has an independent interest in the literature, we have the source solutions

$$U(x,t) = \frac{2t}{At^2 + x^2} \tag{2.23}$$

with $A > 0$ (recall that $n = 1$). Here is the optimal decay estimate:

$$0 \le u(x,t) \le \frac{M^2}{2\pi^2 t}, \qquad M = \int u_0(x)\,dx. \tag{2.24}$$

Note also that $\lim_{m \to -1} c(m, 1) = 0$.

2.2.2 Best constants and optimal problems

The calculation of best constants is a classical topic in the calculus of variations. Typical examples are the best constant C in the Poincaré inequality:

$$\|u - u_\Omega\|_{L^2} \le C\|\nabla u\|_{L^2}$$

among all functions $u \in H^1(\Omega)$, Ω a bounded set in $\mathbf{R}^n$, where u_Ω is the average of u over Ω. A closer subject is the calculation of the best constant in the Sobolev embeddings

$$\|u\|_{L^q(\mathbf{R}^n)} \le C\|\nabla u\|_{L^p(\mathbf{R}^n)}$$

among all functions $u \in W^{1,p}(\mathbf{R}^n)$ with $1 \le p < n$, where $q = pn/(n-p)$, cf. [Au76, Ta76]. In fact, the extremals of that problem for $p = 2$ are the functions

$$W(x,t;a,b,c) = \left(\frac{c}{a^2 + |x-b|^2}\right)^{(n-2)/2}$$

with free constants $a, c > 0$ and $b \in \mathbf{R}^n$, which look precisely like ZKB solutions for $m_c < m < 1$.[2] The deep connection of this family of functions and the Sobolev–Gagliardo–Nirenberg inequalities that they optimize with the smoothing and decay of the PME/FDE has been investigated by del Pino and Dolbeault in [DPD02]. In fact, the connection is even closer when we consider the function $F = W^q$ with exponent $q = 2n/(n-2)$, because this function belongs to L^1 according to the Sobolev embedding. When we look at the extremal function, we land precisely on the Barenblatt profile for $m = (n-1)/n$. This connection has been recently explored by Demange [De05] in his doctoral thesis.

We can view our result in the same optimization light; we state the optimal problem we have solved as: *To find the best upper bound*

$$\sup\{\|u(\cdot,t)\|_\infty : u \in \mathcal{S}_1\}$$

when u runs in the class $\mathcal{S}_1$ of non-negative solutions of the PME with initial data $u_0 \in L^1(\mathbf{R}^n)$ satisfying $u_0 \ge 0$ and $\int u_0(x)\,dx = 1$; $t > 0$ is fixed.

---

[2]Extremals for $p \ne 2$ correspond to Barenblatt solutions of the p-Laplacian equation studied in Chapter 11.

The solution of this problem is just $c(m,n)t^{-\alpha}$ for any $t > 0$ and the optimum is attained by the ZKB solutions. At the same time, we have solved the apparently more general problem: *To find the best upper bound for the quotient*

$$Q(u) = \frac{\|u(\cdot,t)\|_\infty}{\|u_0\|_1^\sigma}$$

when u runs in the class $\mathcal{S}$ of non-negative solutions of the PME with initial data $u_0 \in L^1_+(\mathbf{R}^n)$; $t > 0$ is fixed.

The solution is the same if $\sigma = 2\alpha/n$. We note that the bound is infinity if $\sigma > 0$ is not properly chosen.

2.3 Smoothing exponents and scaling properties

The reader will have observed that the estimates, and the intermediate calculations leading to them, contain a number of exponents that may seem abstruse at first glance. Since, on the other hand, they may have a certain importance in the applications, we would like to be able to predict them in an easy way. This is indeed possible once we realize that they are closely related to the scaling properties of the equation introduced in Section 1.2, cf. formulas (1.22) and (1.23).

Let us show how this property is applied to reduce the proof of existence of the L^1–L^∞ smoothing effect of Theorem 2.1. Indeed, we will prove the following

Proposition 2.2 *The smoothing effect* (2.12) *follows from the case $M = 1$, $t = 1$, i.e., if we are able to prove that for all functions $u_0 \in L^1(\mathbf{R}^n)$ with $u_0 \geq 0$ and $\int u_0\,dx \leq 1$, the following estimate holds:*

$$u(x,1) \leq c. \tag{2.25}$$

Proof We have already seen that PME is invariant under a two-parameter scaling group, so that whenever $u(x,t)$ is a solution of the equation, also $\tilde{u}(x,t) = K\,u(Lx,Tt)$ is a solution, if the K, L and T satisfy

$$K^{m-1}L^2 = T. \tag{2.26}$$

Next, given any solution $u(x,t)$ with data $u_0 \in L^1(\mathbf{R}^n)$, $u_0 \geq 0$ and $\int u_0\,dx = M$, we can choose K, L, T so that $\tilde{u}$ fulfils on top of the above properties the requirement of having L^1-mass 1:

$$\int \tilde{u}_0(x)\,dx = K\int u_0(Lx,0)\,dx = KL^{-n}M.$$

We have two conditions, $K^{m-1}L^2 = T$ and $KM = L^n$, hence

$$L = M^{(m-1)\delta}T^\delta, \quad K = M^{-2\delta}T^{n\delta}$$

with $\delta = 1/(n(m-1)+2)$ and free parameter T. But now, taking $t = 1$ and fixing the free parameter $T > 0$ at will we have

$$\|u(T)\|_\infty = \frac{1}{K}\|\tilde{u}(1)\|_\infty = \frac{c}{K} \le c\, M^{2\delta} T^{-n\delta}.$$

We only have to change the letter T into t to obtain the desired result with constant C, cf. formula (2.12). □

A similar calculation can be done if we know the bound at any other time $t_0 > 0$.

Summing up, the difficulty in the derivation of the decay formula (2.12) does not lie in the exponents attached to mass $M = \|u_0\|_1$ and time t, because these are determined by the scaling properties of the equation, and can be calculated by some simple algebra once we have established that the map $u_0 \mapsto u(t)$ admits a bound for a certain time t_0 and a certain mass $M = \|u_0\|_1$. The real novelty of our result lies therefore in *the existence of a finite constant and the calculation of its precise value*, which we see as an optimization problem. Really minimum effort was needed in doing this part once it was discovered that it is attained by a special solution.

The application of this principle to the rest of smoothing formulas to follow is similar and will often be left to the reader. The general conclusion is that we need only obtain the desired bound for data of norm 1 at time $t = 1$.

We will be strongly using similarity arguments in subsequent chapters, e.g., in Chapter 3 and Section 5.4. A final caveat: the self-similar solutions we will use in the range $m < m_c$ do not have exponents that can be calculated by simple algebra. This is an important issue that will be studied in Chapter 7.

2.4 Strong and weak smoothing effects

In the example of smoothing effect seen so far, we have obtained an estimate of the solution at time t in a 'better space' Y (in this case $L^\infty(\mathbf{R}^n)$) in terms of only the norm of the data u_0 in the original space X (say, $L^1(\mathbf{R}^n)$), and the estimate may also depend on t, on some characteristic of the spaces, and on the structural parameters of the equation, let us call them collectively $\tilde{m}$; but the estimate does not depend on more information about the data or solution. In other words,

$$\|u(t)\|_Y \le F(\|u_0\|_X, t; X, Y, \tilde{m}).$$

Moreover, function F is also non-decreasing in the variable $\|u_0\|_X$. We call such an estimate a *strong smoothing effect* from X into Y. As a consequence of the strong smoothing effect, bounded families of data in X produce bounded families of solutions in Y for every fixed $t > 0$.

For equations with a scaling structure, like the PME, the existence of a strong smoothing effect can be reduced to check it only at time $t = 1$ and for data u_0 of norm equal or less than 1, as we have just shown. This powerful use of the techniques of scaling reduces the effort in calculating the effect.

We will find in the sequel cases where only part of the conclusion is true: for all data in X the solution is in Y for all $t > 0$ (or at least for a time interval), but the dependence function F also includes some other information about the data. This will be called a *weak smoothing effect* from X into Y.

Our main interest lies in finding strong smoothing effects whenever possible. However, examples of weak effects abound. We will discuss some of them: the actual decay of L^p solutions with $p > 1$ for all $m > 0$, treated in Section 3.6; the exponential decay in the critical case $(m, p) = (m_c, 1)$ with $L^1 \cap L^\infty$ data mentioned in Section 5.6; the boundedness of solutions with data in the critical L^p space for $m \geq 3$ of Chapter 6; and finally, the weak L^1–L^∞ effect for the logarithmic diffusion equation discussed in Subsection 8.2.2.

2.5 Comparison for different diffusivities

The estimates of this section can be immediately extended to the filtration equation

$$u_t = \Delta\varphi(u) \tag{2.27}$$

under the condition that $\varphi(u)$ be more diffusive than $\psi(u) = u^m/m$ for some m, see Definition 1.1. Using the comparison result, Theorem 1.3, we conclude that all estimates computed so far are valid with constants equal or less than the ones obtained. Of course, we lose the optimality statement.

We recall that a simple sufficient condition for comparison of diffusivities, consists in asking that

$$\varphi'(s) \geq \psi'(s) = s^{m-1}, \quad \forall s, \tag{2.28}$$

and then φ is more diffusive than ψ.

In fact, there is a variation of the topic that is useful in the applications. We find nonlinearities satisfying

$$\varphi'(s) \geq C\, s^{m-1} \tag{2.29}$$

for given constants $m > 0$ and $C > 0$. In that case we have to modify the filtration equation in order to apply the above comparison. This is done by changing the time scale to absorb the constant. The time transformation is $\tau = k\,t$, with $k = C$, so that

$$u_\tau = \Delta\tilde{\varphi}(u), \quad \tilde{\varphi}(s) = \frac{1}{C}\varphi(s),$$

and $\tilde{\varphi}$ satisfies (2.28). Then the estimates for $u_t = \Delta u^m$ will be true for the filtration equation (2.27) if t is replaced by Ct. Indeed, the solution of equation (2.27) under conditions (2.29) is less concentrated at time t than the solution of the PME at time $\tau = C\,t$.

Exponents, diffusivities, and rates Let us make a comment about the relation between slow/fast diffusion and the decay rates. Recall the warning about the problem with the names slow/fast at the end of Subsection 1.2.1. That remark has a confirmation in the variation with m of the exponent rates of the L^1–L^∞ effect. The rate of decay of the Barenblatt solutions is $U = O(t^{-\alpha})$, where

$$\alpha = \frac{n}{n(m-1)+2} = \frac{1}{m-m_c}$$

is a monotone decreasing function of m. For more general L^1-data, the rate is exact only for large times, see Section 2.8 below. Note that in the heat equation the exponent is $n/2$ (with space expansion rate $\beta = 1/2$, corresponding to standard Brownian motion). When m varies from m_c to ∞, the exponent varies from ∞ to 0.

Here is an explanation of the relationship: the solution decays faster for smaller m when t is a very large so that u is already small. However, near $t = 0$, when L^1 solutions may be large, the bound on the decay is faster when $m \to \infty$. A similar comment applies to the spread rate $\beta = \alpha/n$.

These considerations also apply to the results of the sections to follow.

2.6 A general smoothing result

At this point, it is necessary to mention the important smoothing result of Bénilan and Berger [BB85] for the Cauchy problem for the filtration equation (2). Assuming that the integral

$$\Phi(k) = \int_k^\infty \frac{ds}{\varphi(s)^{n/(n-2)}}, \qquad s > 0 \tag{2.30}$$

converges, they prove that there exists a function F such that

$$\|\varphi(u(t))\|_{L^\infty(\mathbf{R}^n)} \le (1+\lambda)\, F\left(C_N \|u_0\|_{L^1(\mathbf{R}^n)}^{2/(n-2)} \left(\frac{\lambda t}{1+\lambda}\right)^{n/(n-2)} \right). \tag{2.31}$$

The function F is defined as $\varphi \circ \Phi^{-1}$, and λ is any positive constant. We will not enter into the proof of this important result that establishes a strong smoothing effect for more general filtration equations and gives the correct decay exponents when applied to the PME, but avoids the discussion of best constants. Note: we have simplified the statement a bit for convenience.

2.7 Estimating the smoothing effect into L^p

After the strong result of the smoothing result from $L^1(\mathbf{R}^n)$ (or the space of bounded measures) into $L^\infty(\mathbf{R}^n)$, we continue with a light section which extends and comments on the strong results of the previous one in the form of smoothing effects into $L^p(\mathbf{R}^n)$ with $1 \le p < \infty$, which is an easy extension.

Corollary 2.3 *In the above situation, for every $u_0 \in L^1(\mathbf{R}^n)$, and every $p \in (1,\infty)$, we have $u(t) \in L^p(\mathbf{R}^n)$ and*

$$\|u(\cdot,t)\|_p \le c(m,n,1,p)\, \|u_0\|_1^{\sigma(1,p)} t^{-\alpha(1,p)} \tag{2.32}$$

with

$$\alpha(1,p)=\frac{n(p-1)}{(n(m-1)+2)p},\qquad \sigma(1,p)=\frac{n(m-1)+2p}{(n(m-1)+2)p}. \tag{2.33}$$

Initial data can also be measures and the result holds.

Proof (i) Finding that there is a constant and the correct exponents is quite standard, using the case $p=\infty$ already proved and the property of conservation of the L^1 norm: $\|u(t)\|_1=\|u(0)\|_1=M$, which amounts to say that $\alpha(1,1)=0$ and $\sigma(1,1)=1$ with $c(m,n,1,1)=1$ (which cannot be improved).

We show a simple derivation of the case $1<p<\infty$ for future reference. We consider in a first step data $u_0\in L^1(\mathbf{R}^n)\cap L^p(\mathbf{R}^n)$. For every $t>0$ we have

$$\|u(t)\|_p^p\le\|u(t)\|_1\|u(t)\|_\infty^{p-1}.$$

Using the a priori bounds on $\|u(t)\|_1$ and $\|u(t)\|_\infty$ we get

$$\|u(t)\|_p^p\le c(m,n,1,\infty)^{p-1}\,M^{1+(p-1)\sigma_1}t^{-(p-1)\alpha_1},$$

where $\alpha_1=\sigma(m,n,1,\infty)$, $\sigma_1=\sigma(m,n,1,\infty)$, are the decay exponents for $p=\infty$ given in Theorem 2.1. The result follows with a constant

$$c=c(m,n,1,\infty)^{(p-1)/p},$$

which is not supposed to be optimal.

(ii) In fact, the optimal constant $c(m,n,1,p)$ is again obtained by performing the computation on the ZKB solutions, since the concentration relation implies ordering of the L^p norms, cf. Lemma 1.1. □

This smoothing effect extends without changes to the range $-1<m\le 0$ in dimension $n=1$.

2.8 Asymptotic sharpness of the estimates

The smoothing estimate (2.12), its companion (2.22), and (2.32), are sharp in the sense that the exponents of the formula are exact and even determined a priori, and the constant is best for the class of functions under consideration, i.e., non-negative integrable functions with mass equal or less than M. But the formula is sharp in a deeper sense: it is sharp for every individual solution with finite mass as time goes to infinity. Indeed, given a function $u_0\ge 0$ in $L^1(\mathbf{R}^n)$ with mass $M>0$ the constant $c(m,n,1,\infty)$ in estimate (2.12) cannot be improved for this particular trajectory $u(t)$. The same result applies to the smoothing effect into $L^p(\mathbf{R}^n)$.

Proposition 2.4 *Let $m>m_c$ and let u be a solution with data $u_0\in L^1(\mathbf{R}^n)$, with $u_0\ge 0$ and $\int u_0(x)\,dx=M$. Then, for every $p\in[1,\infty]$ we have*

$$\lim_{t\to\infty}\|u(t)\|_p\,M^{-\sigma(1,p)}t^{\alpha(1,p)}=c(m,n,1,p). \tag{2.34}$$

Proof We first notice that formula (2.34) is exact for the ZKB solutions at all times, not only as a limit. When u is a general solution the result is an easy consequence of the previous fact and the result on asymptotic convergence of any solution towards the ZKB solution with the same mass $U(t)$, which is proved in [Va03, Theorem 1.1]. We give the full statement of the result on the asymptotic behaviour of the solutions of the PME with L^1 data for the reader's convenience.

Theorem 2.5 *Let* $u(x,t)$ *be the unique weak solution of the Cauchy problem for the PME with initial data* $u_0 \in L^1(\mathbf{R}^n)$, $u_0 \geq 0$. *Let* U_M *be the Barenblatt solution with the same mass as* u_0. *Then as* $t \to \infty$ *we have*

$$\lim_{t\to\infty} \|u(t) - U_M(t)\|_1 = 0. \tag{2.35}$$

Convergence holds also in L^∞*-norm in the proper scale:*

$$\lim_{t\to\infty} t^{\alpha} \|u(t) - U_M(t)\|_\infty = 0 \tag{2.36}$$

with $\alpha = n/(n(m-1)+2)$. *Moreover, for every* $p \in (1,\infty)$ *we have*

$$\lim_{t\to\infty} t^{\alpha(p)} \|u(t) - U_M(t)\|_{L^p(\mathbf{R}^n)} = 0 \tag{2.37}$$

with $\alpha(p) = \alpha(p-1)/p$.

The conclusion is that given a particular solution, even if we take into account more information on the data $u_0 \in L^1(\mathbf{R}^n)$, there is no possible smoothing effect of the weak type that provides more accurate information and is also valid for large times. A proof of the robustness of the result is the fact that it still holds for signed solutions on the condition that the total mass $M = \int u_0(x)\,dx \neq 0$. For solutions of zero mass se comment below. □

The situation will be different for other smoothing effects to appear soon.

Convergence with error rates A very interesting question is the calculation of the so-called convergence rates, i.e., the rate at which the right-hand-side of formulas (2.35)–(2.37) goes to zero. This subject has been the object of much recent research using methods which combine the study the evolution of certain functionals called *entropies* with functional inequalities that are related to generalized Sobolev inequalities. We refrain from going into such a deep subject and recommend as starting references [CT00, CJMU, DPD02, Ot01]. Let us only add in this context that the proof of existence of decay rates in these works is completely different, it relies on more delicate arguments, but on the other hand, it gives a more precise information in the form of convergence to asymptotic profiles with an extra decay rate for the error.

2.9 The limit $m \to \infty$. Mesa problem

We have already commented on the limit of the estimates of Section 2.2 in the lower end when $m \to m_c$. There is another limit situation, namely letting $m \to \infty$. In that case we see that

$$\alpha \to 0, \qquad \sigma \to 0.$$

With a bit more of effort we can check in formula (2.13) that

$$\lim_{m\to\infty} c(m,n) = 1.$$

The following question can be posed: is there a meaning to these formulas? Actually, there is, it has been carefully investigated, and it introduces some interesting ideas. Here is a brief account of the issue.

Suppose that we take fixed initial data $u_0 \in L^1(\mathbf{R}^n)$, $u_0 \geq 0$ and we solve the PME for arbitrary $m > 1$. We find a family of solutions $u_m(x,t) \in C([0,\infty) : L^1(\mathbf{R}^n))$. The smoothing estimates hold, so the family is uniformly bounded for all $t \geq \tau > 0$. We may thus pass to the limit (at least in a weak sense) and find a function

$$u_\infty(x,t) = \lim_{m\to\infty} u_m(x,t) \tag{2.38}$$

where the limit is taken at least along subsequences $m_k \to \infty$. Our a priori estimates hold for u_m, hence we conclude in the limit that

$$0 \leq u_\infty(x,t) \leq 1 \tag{2.39}$$

for all $x \in \mathbf{R}^n$ if $t > 0$. We conclude from this that for general initial data there must be a discontinuity between the value at $t = 0$ and the values for $t > 0$, and we cannot have the naively expected regularity $u_\infty \in C([0,\infty) : L^1(\mathbf{R}^n))$.

The explanation of the universal bound $u_\infty(x,t) \leq 1$ is quite clear in physical terms. The diffusivity of the elliptic operator is $D(u) = u^{m-1}$. This means in the range $m \approx \infty$ a very high diffusivity at all points where $u > 1$, hence all the mass located there tends to be quickly diffused away into regions of lesser concentration u; in the limit, no such region persists and u becomes less than 1. On the other hand, regions where $u_0 < 1$ tend not to be affected by the diffusive process since $D(u) \to 0$ as $m \to \infty$. However, we have to take into account the effect of the collapse on the lower concentrations of the mass initially associated with high concentrations.

Actually, it can be proved that u_m converges uniformly for $t \geq \tau > 0$ and all $x \in \mathbf{R}^n$, cf. [BBH] to a unique limit $u_\infty(x,t)$. The discontinuity at $t = 0+$ is called an *initial discontinuity layer*. Moreover, the final profile is independent of time

$$u_\infty(x,t) = F(x). \tag{2.40}$$

This profile is called in the technical literature a *mesa*, because of its resemblance with such features of the landscape of the far West.

In case $u_0 \leq 1$, then $F(x) = u_0(x)$. Otherwise, calculating the stationary profile F in terms of u_0 implies solving a variational inequality that projects the data u_0 into a function with values in the interval $[0, 1]$ in the following way [CF87]: there is a set $A \subset \mathbf{R}^n$ such that $F(x) = 1$ if $x \in A$, $F(x) = u_0(x)$ if $x \notin A$. Moreover, A is the non-coincidence set of the variational inequality:

$$\Delta w + f \leq 1, \quad w \geq 0, \quad (\Delta w + f - 1)w = 0 \tag{2.41}$$

with $f = u_0$. This limit plays for $m = \infty$ a role equivalent to the ZKB asymptotic profiles for $m < 1$. There are two striking differences:

(i) the family of possible asymptotical models $F(u_0)$ is not a countable family,
(ii) stabilization takes places in zero time!

Regarding the latter aspect, an important question is estimating the width of the initial layer. This is done by estimating the time τ a typical solution with data in $L^1(\mathbf{R}^n)$ takes to be become bounded, say

$$u(x,t) \le K \quad \text{for} \quad x \in \mathbf{R}^n,\ t \ge \tau.$$

Of course $\tau = \tau(m, n, K)$. We refer to Exercise 2.1 for some explicit calculations. The question has been investigated in [FH].

2.10 Comments and historical notes

Section 2.1 The source solution for the PME was introduced by Zeldovich and Kompanyeets [ZK50] in a particular case and completely analysed by Barenblatt [Ba52] in the period 1950–52. It was independently introduced in the West by Pattle in [Pa59] in 1958. Its role in describing the asymptotic behaviour will be commented upon below.

The algebraic decay rate (in space) of the ZKB in the fast diffusion range gets slower and slower as m goes down, and stops being integrable at $m = m_c$, precisely the value for which the ZKB ceases to exist in the integrable sense, as the evolution of a point source, though it still has a meaningful life of its own that is explored in Exercise 5.9 where we use algebraic continuation.[3]

Thus, the ZKB family of point source solutions disappears at the critical exponent $m = m_c$. Then a relative can be found but it is not an integrable distribution, see Example 5.1, formula (5.37). For $m < m_c$ the algebraic formula can be continued into solution (5.2) which models extinction, and the larger family constructed in Exercise 5.2; but the role is not the same and maybe the closest relative in behaviour is the family of self-similar solutions of the second kind constructed in Chapter 7. If we see the ZKB solutions from the point of view of source solutions, then for $m \le m_c$ there is no solution in the standard sense of the Cauchy problem with a Dirac delta as initial data [BF83]: more generally, existence is reduced to initial measures satisfying some capacity conditions [Pi87].

Regarding the fat tails in fast diffusion, the existence of overpopulated distributions is a popular topic in kinetic equations. Thus, Baldasarri et al. [BMP02] performed Monte Carlo simulations of the spatially homogeneous Boltzmann equation for inelastic Maxwell molecules, and have thus shown that general classes of initial distributions

[3] Algebraic continuation is a very powerful tool when available, but it has rather unpredictable implications for other points of view.

evolve for large times into a singular nonlinear scaling solution with a power law tail. This is precisely the case of our fast diffusion tails.[4]

The topic of the Lagrangian approach to diffusion is discussed in Appendix II. It allows us to talk of mass distribution versus velocity distribution. A very curious property of the ZKB solutions is proved; *if we assume that the mass distribution obeys a ZKB profile, also the velocity distribution is a ZKB profile.* To be noted the quite simple linear relation between velocities and space location typical of the ZKB solutions: $\mathbf{v} = \beta x/t$. Note also that β increases as m decreases.

The strongly singular solution (2.11) is studied in [ChV02] where it is called IPSS (infinite point source solution) and is used to construct a theory of singular solutions called *extended continuous solutions* which allow for hot spots where the temperature $u = \infty$ at all times. Well-posedness of the Cauchy problem can be proved in a setting that extends the theory of weak solutions. The role of the IPSS in the asymptotics is carefully discussed.

Section 2.2 The semigroup approach was used in the 1970s by Bénilan [Be76] and Véron [Ve79] to derive smoothing effects and decay rates. Our contribution with respect to them in this section lies in obtaining the best constants, and also in the simplicity of the worst-case strategy proposed in [Va82] that can be easily generalized to other problems. In taking this approach we were inspired by Talenti's work on Sobolev inequalities [Ta76]; see also Aubin [Au76].

Actually, the question of best constants has recently recovered attention from researchers because of the connection between the decay in parabolic problems and certain kinds of Sobolev inequalities that we briefly mention in Subsection 2.2.2, see more in [DPD02, CT00]. Useful references treating the connection with the heat kernel are the books by Davies and Saloff-Coste [Dv89, SC01]. The problem of best constants in Sobolev inequalities on manifolds is addressed by Hebey in [Hb99]. For the extension to p-Laplacian equation see [DDG04].

This text will use repeatedly the strategy of comparing with a worst case that happens to be a self-similar solution. This is no coincidence, since it is commonplace in mathematical physics that self-similar solutions provide the asymptotic representation of a wide range of important phenomena when the underlying equations have some sort of scaling properties, either exact or approximate. This fact has been confirmed both in theory and by the numerical experience. See [Ba79] for an account of the subject.

We stress again the fact that the exponent restriction $m > m_c$ is essential in the smoothing effect stated here. The surprising properties of subcritical fast diffusion are studied in detail in Chapter 5 *et seq.*

Section 2.3 The use of scaling properties is usually referred to in the physics literature as dimensional analysis, cf. e.g., [Ba87]. The relation between scaling and self-similarity is investigated in detail in Chapter 16 of [V06].

---

[4] I thank G. Toscani for information on this subject.

Section 2.4 The distinction between strong smoothing effects and weak smoothing effects plays an important role in the sequel. We have not found it systematically treated in the existing literature.

Section 2.6 The existence of a smoothing effect from $L^1(\mathbf{R}^n)$ into $L^\infty(\mathbf{R}^n)$ for the filtration equation has been characterized by Bénilan and Berger [BB85] as equivalent to the convergence of integral (2.30).

Section 2.8 The precise asymptotic behaviour of finite mass solutions of the PME towards the Barenblatt solutions is due to Kamin [Km73] in dimension $d = 1$ and then to Friedman and Kamin [FK80] in several dimensions; their proof works for solutions with compactly supported initial data, otherwise the convergence is local on expanding sets. Improvements were done in [KV88]. The complete proof of convergence in all $L^p(\mathbf{R}^n)$, $1 \le p \le \infty$, as stated in this section appeared in [Va03].

These works deal mainly with $m > 1$ but the arguments extend to the whole range $m > m_c$. In the fast diffusion range $m_c < m < 1$ a more precise result of convergence in relative error is proved in Theorem 21.1 of [Va03].

The result for signed solutions is due to [KV91] when the total mass does not vanish. The asymptotic rates and profiles have been studied in [KV91] in the case of zero mass in $n = 1$ and a dipole solution is found as the asymptotic model. The work is extended to several dimensions in [HuV91] by constructing a several-dimensional dipole.

For a detailed analysis of the asymptotic behaviour of the PME/FDE in different settings the reader is referred to [V06], Chapters 18–20.

Section 2.9 The first mathematical reference to the mesa problem for the PME seems to be [E4-86]. We refer to [CF87, Sk89, FH] for important progress on the issue of convergence and identification of the limit. A more general consideration on singular limits appears in [BEG].

The results we present here are not the only way in which we can pass to the limit as $m \to \infty$. Actually, it makes much sense to perform the limit when the initial pressure $v_0 = u_0^{m-1}/(m-1)$ is kept fixed. This is a limit in another scale and it leads to quite different and interesting results when boundary data are prescribed in a domain which is a subset of $\mathbf{R}^n$, see [AGV98, GQ03].

We will study in some detail another case of initial layer formation, namely the case when $m \to -1$ (the lower end limit) in dimension $n = 1$, cf. Section 9.4.1. In that case, the solution collapses to zero across the initial layer.

Exercises

Exercise 2.1 (i) Calculate the limit of the ZKB solutions as $m \to \infty$. Show that it is a mesa of constant height 1 with radius $R = d\,\|u_0\|_1^{1/n}$. Calculate d by using conservation of mass.

(ii) Use formula (2.15) to prove that for $n = 2$ the width of the initial layer can be estimated as

$$\tau(m, 2, C) = \frac{1}{4\pi} K^{-m}.$$

(iii) Use formula (2.13) to prove that for $n \geq 1$ the width of the initial layer can be estimated in first approximation by the same formula with $1/4\pi$ replaced by a certain constant

$$c_\infty = \lim_{m\to\infty} c(m, n)^m$$

(iv) Conclude that the width of the initial layer decreases approximately linearly with $1/\|u(t)\|_\infty^m$ for all $m \gg 1$. A convenient variable in this respect seems to be u^m and not u.

Exercise 2.2 (i) Prove that when $0 \leq u_0 \leq 1$ the mesa profile $F(u_0)$ described in Section 2.9 in the limit $m \to \infty$ is just $F = u_0$.

(ii) Derive the variational inequality that allows us to project u_0 into F for general u_0 (see cited literature).

Exercise 2.3 Note that for the embeddings L^1–L^p we have

$$\lim_{m\to\infty} \alpha(m, n; 1, p) = 0, \qquad \lim_{m\to\infty} \sigma(m, n; 1, p) = \frac{1}{p}.$$

Explain those numbers by interpolation.

Exercise 2.4 According to the Lagrangian formulation the speed of the particles that are supposed to move according to the PME flow is given by

$$\mathbf{v} = -u^{m-2}\nabla u = -\nabla v, \qquad v = \frac{1}{m-1} u^{m-1}$$

(using the modified equation; in the old form the pressure is $v = mu^{m-1}/(m-1)$). We want to explore the consequences of assuming a linear dependence

$$\mathbf{v} = \beta(t)x,$$

for some scalar β function of time. This generalizes what is found for the ZKB solutions where $\beta(t) = \beta/t$.

(i) Show that this assumption leads to a pressure of the form

$$v(x, t) = A(t)x^2 + B(t). \tag{2.42}$$

(ii) Derive the pressure equation

$$v_t = v\,\Delta v + \gamma\,|\nabla v|^2 \quad \text{with } \gamma = 1/(m-1). \tag{2.43}$$

(iii) Using this equation calculate all the solutions of the PME of the form (2.42). Recover the ZKB solutions. Describe the rest of the solutions.

Exercise 2.5 (i) Exponent $m = (n-1)/n$ (in dimension $n \geq 2$) is important in connection with Sobolev inequalities mentioned in Subsection 2.2.2. Calculate the constants of the smoothing effect (2.12) in that case and show that $\alpha = n$, $\sigma = 2$ and

$$c(m,n)^{1/2n} = \frac{1}{\sqrt{2n\pi}} \left\{ \frac{\Gamma(n)}{\Gamma(n/2)} \right\}^{1/n}.$$

This constant is to be compared with the value obtained for the best constant of the Sobolev embedding $W^{1,2}$ into $L^{2n/(n-2)}$, $n > 2$, which is:

$$C = \frac{1}{\sqrt{n(n-2)\pi}} \left\{ \frac{\Gamma(n)}{\Gamma(n/2)} \right\}^{1/n}.$$

(ii) Calculate the constants of the smoothing effect (2.12) for $m = n/(n-2)$ with $n > 2$. Show that

$$\alpha = \frac{n(n-2)}{4(n-1)}, \qquad \sigma = \frac{n-2}{2(n-1)},$$

and

$$c(m,n)^{1/2\alpha} = \frac{1}{\sqrt{4(n-1)\pi}} \left\{ \frac{\Gamma(n)}{\Gamma(n/2)} \right\}^{1/n}.$$

Taking as equation $u_t = \Delta u^m$, the best constant is such that

$$(c'(m,n))^{1/2\alpha} = \frac{c(m,n)^{1/2\alpha}}{m^{1/2}} = \frac{1}{\sqrt{n(n-2)\pi}} \left\{ \frac{\Gamma(n)}{\Gamma(n/2)} \right\}^{1/n},$$

which is precisely the Aubin–Talenti constant.

Exercise 2.6 The limits $m \to 1$. We know that the density u_m of the solution of the PME with exponent $m > 1$ converges to the solution of the heat equation in the limit $m \to 1$ when the initial data are kept fixed

$$u_m(x,0) = \phi(x) \in L^1(\mathbf{R}^n), \quad u_0 \geq 0.$$

We have already mentioned how the smoothing estimates pass to the limit. A second way of passing to the limit $m \to 1$ is possible in which the initial pressure is kept fixed

$$v_m(x,0) = \psi(x) \in L^1(\mathbf{R}^n), \quad v_0 \geq 0$$

where, as already explained, $v_m = u_m^{m-1}/(m-1)$. This is a singular limit for the density variable. The equation satisfied by v_m is

$$v_t = (m-1)v,\, \Delta v + |\nabla v|^2,$$

so that in the limit $m \to 1$ we expect v_m to converge to the solution of the eikonal equation

$$v_t = |\nabla v|^2$$

with the same data. This has been rigorously proved in [AV87, LSV87]. Pass to the limit $m \to 1$ in the smoothing estimates and explain their meaning for the eikonal equation. Do we get a smoothing effect?

3

Smoothing effect and time decay from L^p or M^p

We now address the question of boundedness for the same equation when the initial data are chosen in the Lebesgue space L^p, $p \in (1, \infty)$. We still assume that $m > m_c$. The results are based on a very delicate phase-plane analysis of the existence of certain types of self-similar solutions. Since this technique will play a big role in later chapters, we gather such analysis as an independent section, Section 3.2, for easier reference. The technique leads to extending the functional setting in a natural way from the Lebesgue spaces into the Marcinkiewicz spaces $M^p(\mathbf{R}^n)$.

Using this machinery, we construct a special solution in Section 3.3 that replaces the ZKB in the present context and allows us to establish the smoothing effect from M^p into L^∞ in Section 3.4. This effect is easily extended into a similar effect that takes place into L^q for all $q > p$. Moreover, self-similar examples show that smoothing effects cannot exist from $M^p(\mathbf{R}^n)$ into $M^q(\mathbf{R}^n)$ with $1 \le q < p \le \infty$. This is covered in Section 3.5, where the general smoothing result is stated, Theorem 3.6.

To complete the picture, we devote Section 3.6 to discuss the weak estimates that are obtained for the smoothing L^p–L^p. For completeness, the case $0 < p < 1$ is examined and the conclusion about existence of smoothing effects is negative.

Finally, we recall that the preceding estimates are global in the sense that they involve L^p norms in the whole space $x \in \mathbf{R}^n$. However, fast diffusion equations are characterized by having very powerful local estimates. This important subject is introduced in Section 3.8.

Sections 3.1–3.5 are recommended for a basic reading, and then Sections 3.6 and 3.8. Section 3.2 is essential for the sequel.

3.1 Strong smoothing effect

In view of last chapter and the general considerations, the reader will not be surprised to find the statement of our next result.

Theorem 3.1 *For every $u_0 \in L^p(\mathbf{R}^n)$, $1 < p \le \infty$, and every $t > 0$ we have $u(t) \in L^\infty(\mathbf{R}^n)$ and*

$$|u(x,t)| \le c\, \|u_0\|_p^{\sigma_p} t^{-\alpha_p} \tag{3.1}$$

with suitable exponents α_p and σ_p and a best constant $c = c(m, n, p, \infty) > 0$.

Proof Using the scaling techniques as described in Section 2.3, it is not difficult to prove that the only possible exponents are a priori given by

$$\alpha_p = \frac{n}{n(m-1)+2p}, \qquad \sigma_p = \frac{2p}{n(m-1)+2p}, \tag{3.2}$$

so the only remaining task is proving that there exists a finite constant and to determine whether it is attained or not. Though this question could look like a mere extension of the preceding results, it offers several new and interesting perspectives. In particular, Marcinkiewicz spaces appear as the natural setting for the estimates.

Before we proceed with the extension, let us prove the present result as a consequence of the L^1–L^∞ case and the comparison of equations with different diffusivities (see Section 2.5). We assume $m \geq 1$. Take a solution u with data in $u_0 \in L^p(\mathbf{R}^n)$, take $\varepsilon > 0$, and consider the function $v_\varepsilon = u - \varepsilon$. This function is the solution of a filtration equation (FE$_\varepsilon$): $v_t = \Delta\Phi_\varepsilon(v)$ with

$$\Phi_\varepsilon(s) = \frac{1}{m}\left((s+\varepsilon)^m - \varepsilon^m\right).$$

Clearly, for every $s \geq 0$ we have $\Phi'_\varepsilon(s) = (s+\varepsilon)^{m-1} \geq \Phi'_0(s)$. This means that for non-negative values, the new diffusivity is larger than the original one, hence by the comparison result, Theorem1.3, the L^∞-norm of non-negative solutions goes down if the initial data are kept and ε is increased.

Since v_ε has changing sign, we need to consider the solution $\tilde{v}_\varepsilon(x,t)$ of $v_t = \Delta\Phi_\varepsilon(v)$ with initial data

$$\tilde{v}_\varepsilon(x,0) = (u_0(x) - \varepsilon)_+ \geq v_\varepsilon(x,0).$$

We have $\tilde{v}_{0,\varepsilon} \in L^1(\mathbf{R}^n)$. Moreover, since $\|u_0\|_p^p \geq \varepsilon^p \operatorname{meas}(\{u_0 \geq \varepsilon\})$, we get

$$\|\tilde{v}_{0,\varepsilon}\|_1 = \|(u_0-\varepsilon)_+\|_1 \leq \|u_0\|_p |\{u_0 \geq \varepsilon\}|^{(p-1)/p} \leq \frac{\|u_0\|_p^p}{\varepsilon^{p-1}}. \tag{3.3}$$

The comparison of diffusivities implies that $\|\tilde{v}_\varepsilon(t)\|_\infty \leq \|U(t)\|_\infty$, where $U(x,t)$ is the solution of the PME having as initial data the symmetrization of $\tilde{v}_\varepsilon(x,0)$. Since the smoothing effect has been proved in that case, see Theorem 2.1, we get

$$\|\tilde{v}_\varepsilon(t)\|_1 \leq c(m,n)\|\tilde{v}_{0,\varepsilon}\|_1^{\sigma'} t^{-\alpha'} \leq c(m,n)\|u_0\|_p^{p\sigma'} t^{-\alpha'} \varepsilon^{-(p-1)\sigma'},$$

where we have used formula (3.3) and the notation $\alpha' = \alpha(1,\infty)$ and $\sigma' = \sigma(1,\infty)$.

We next observe that $v(x,t) = u(x,t) - \varepsilon \leq \tilde{v}_\varepsilon(x,t)$ for every $x \in \mathbf{R}^n$ and $t > 0$, by the standard comparison theorem for solutions of the filtration equation (FE$_\varepsilon$). Therefore,

$$\|u(t)\|_\infty \leq \varepsilon + c(m,n)\|u_0\|_p^{p\sigma'} t^{-\alpha'} \varepsilon^{-(p-1)\sigma'}.$$

We can minimize this expression in ε since $\varepsilon > 0$ was taken arbitrary. This gives the desired result with correct exponents,

$$\alpha = \frac{p\sigma'}{1+(p-1)\sigma'}, \qquad \sigma = \frac{\alpha'}{1+(p-1)\sigma'},$$

that agree with (3.2), and some constant that need not be the optimal constant $c(m, n, p, \infty)$. □

Remark on the mesa limit When we consider the embeddings into L^p, $p < \infty$, we have

$$\lim_{m\to\infty} \alpha(m, n; 1, p) = 0, \qquad \lim_{m\to\infty} \sigma(m, n; 1, p) = \frac{1}{p}.$$

3.2 Scaling and self-similarity

The use of scaling methods, and the self-similar solutions that naturally originate from it, will be a quite important tool in many parts of the this text. Therefore, we find it convenient to introduce in this moment the main results on self-similar methods applied to the PME/FDE, as a kind of specialized appendix for readers who are not familiar with the subject. Subsections 3.2.1, 3.2.2 are required reading, while 3.2.3 and 3.2.4 can be skipped until needed.

3.2.1 Scaling and self-similar solutions

Let us apply the ideas of scaling to obtain self-similar solutions for scale invariant data. Assume that the initial data have the form $u_0(x) = A\,|x|^{-\gamma}$. Assume also that for certain values of $\gamma \in \mathbf{R}$ we have an existence and uniqueness theorem in a class of solutions. Let $u = u(x, t)$ be the solution for such data. The scaling transformation (1.22)–(1.23) is now applied to u to produce another solution $\tilde{u}$. Now, if $K = L^\gamma$ we have as initial data

$$\tilde{u}(x, 0) = KA\,|Lx|^{-\gamma} = u(x, 0).$$

By uniqueness of solutions, we conclude that $u(x, t) = \tilde{u}(x, t)$, i.e.,

$$u(x, t) = \tilde{u}(x, t) = K\, u(Lx, Tt).$$

Since the scaling group operates under the conditions $K^{m-1}L^2 = T$, we finally have

$$K = T^{\gamma\beta}, \quad L = T^{\beta},$$

with $\beta = 1/(\gamma(m-1)+2)$, which is well-defined if

$$\gamma \neq -2/(m-1).$$

We will assume that restriction in what follows. Fix now $t_0 > 0$ and put $T = 1/t_0$. We get the formula $u(x, t_0) = t_0^{-\alpha} f(x\, t_0^{-\beta})$, with $\alpha = \gamma\beta$ and $f(y) = u(y, 1)$. Since $t_0 > 0$ is arbitrary this can be written in the self-similar form

$$u(x,t) = t^{-\alpha} f(|x|\, t^{-\beta}), \tag{3.4}$$

and f is called the *self-similar profile*; α and β are called *similarity exponents*. Solutions of this form are called self-similar solutions.

The argument also shows that the solution must live forever. Suppose for instance that u were only defined in a finite time interval $0 < t < T_1$. Fix now $T < 1$ and define K, L and $\tilde{u}$ as before. The latter is then defined for $0 < t < T_1/T$. But the argument shows then that $u = \tilde{u}$ in the common time interval $0 < t < T_1$. This means that $\tilde{u}$ extends u to the interval $0 < t < T_1/T$. Since T is arbitrary, we may perform the maximal extension and take $T_1 = \infty$.

The reader will have noticed that only general information is used from the equation: existence and uniqueness for certain data (and only locally in time), and existence of a scaling transformation. Therefore, the argument applies not only to the PME and the FDE, but also to the p-Laplacian equation and many other equations, not necessarily parabolic. Note also that the scaling group algebra is different for different equations, hence the difference in the obtained similarity exponents.

In Section 7.1.1 we will introduce a different type of similarity formulas that exist only for a finite time in the future. For the sake of distinguishing them, the present form, $U(x,t) = t^{-\alpha} f(|x|\, t^{-\beta})$, will be referred to as *Type-I self-similarity*.

3.2.2 Derivation of the phase-plane system

We want to find solutions in self-similar form

$$U(x,t) = t^{-\alpha} f(\xi), \quad \xi = |x|\, t^{-\beta}, \tag{3.5}$$

for the PME/FDE equation written in the form

$$u_t = \Delta(u^m/m), \quad m > 0. \tag{3.6}$$

We get the relation $\alpha(m-1) + 2\beta = 1$ and the ODE for f

$$\xi^{1-n}(\xi^{n-1} f(\xi)^{m-1} f'(\xi))' + \beta\xi\, f'(\xi) + \alpha\, f(\xi) = 0. \tag{3.7}$$

Now comes the interesting point of this method. We introduce the following variables:

$$\xi = e^r, \; X(r) = \frac{\xi f'}{f}, \; \text{and } Y(r) = \xi^2 f^{1-m}. \tag{3.8}$$

We will prove that the functions $X(r)$ and $Y(r)$ satisfy the following autonomous system:

$$\begin{cases} \dot{X} = (2-n)X - mX^2 - (\alpha + \beta X)Y, \\ \dot{Y} = (2 + (1-m)X)Y, \end{cases} \tag{3.9}$$

where $\dot{X} = dX/dr$, $\dot{Y} = dY/dr$. The second line is immediate from the definition of Y. The derivation of the first line is as follows: the first term of (3.7) gives

$$\xi^{1-n}(\xi^{n-1} f^{m-1} f'(\xi))' = \xi^{1-n}(\xi^{n-2} f^m X)' = \xi^{1-n}(\xi^n f\, X/Y)'.$$

Working it out we get

$$\frac{\dot{X}}{Y}f - \frac{X}{Y^2}\dot{Y}f + \frac{X^2}{Y}f + n\frac{X}{Y}f.$$

The whole equation then reads

$$\frac{\dot{X}}{Y}f - \frac{X}{Y^2}\dot{Y}f + \frac{X^2}{Y}f + n\frac{X}{Y}f + \beta X f + \alpha f = 0.$$

Use now $\dot{Y} = (2 + (1-m)X)Y$ to get

$$\dot{X} - (2 + (1-m)X)X + X^2 + n\,X + (\beta X + \alpha)Y = 0.$$

This completes the derivation of system (3.9). The analysis of this system is easy by phase plane techniques and gives a very detailed knowledge of the class of radially symmetric, self-similar solutions of the PME/FDE equation. Note some of the most important characteristics.

Critical points There are at most three critical points. Indeed, second line of system (3.9) selects the values $Y = 0$ and $X = -2/(1-m)$. For $Y = 0$ there exist two critical points:

$$O = (0, 0), \qquad A = ((2-n)/m, 0),$$

the second being defined if $m \neq 0$. It is different from O if $n \neq 2$.

The other option is the point $B = (X_B, Y_B)$ with the choice $X_B = 2/(m-1)$ given by the second line. Then, the first line vanishes at the isocline of vertical slopes

$$Y = -\frac{X(mX + n - 2)}{\beta(\gamma + X)},$$

which for $X = X_B$ gives

$$Y_B = -X_B(2 + n(m-1)) = -\frac{2\,(2 + n(m-1))}{m-1}.$$

The point B is defined for $m \neq 1$ and coincides with A for $m = m_c$, unless $n = 2$, $m_c = 0$, when A is not defined.

(i) In this book, we use such an information in the study of the forward and backward smoothing effects with data of the form $u_0(x) = A\,|x|^{-\gamma}$ with $\gamma > 0$. As the profile is positive, we only need to consider orbits where $\{Y > 0\}$. Since it is monotone decreasing, we have $X < 0$. All together, we work in the quadrant

$$Q = \{X < 0,\ Y > 0\}.$$

As for the end values, there are two options: (i) if $\alpha > 0$ we have to look at the limit as $\xi \to \infty$ to recover the initial situation, i.e., as $r \to \infty$. We will have $f(\xi) \sim A\,\xi^{-\gamma}$ as $\xi \to \infty$, hence

$$X(\infty) = -\gamma, \quad Y(r) \sim \xi^{-\gamma(1-m)+2}$$

as $r \to \infty$, and the latter goes to ∞ since $\beta > 0$. For our conclusions about regularity, we need to know what happens as $r \to -\infty$ in order to derive the behaviour of $F(\xi)$ for $\xi \sim 0$.

(ii) On the contrary, when $\alpha < 0$, we have to look at the limit as $\xi \to 0$, i.e., $r \to -\infty$ and we will have $f(\xi) \sim A\,\xi^{-\gamma}$, hence

$$X(-\infty) = -\gamma, \quad Y(r) \sim \xi^{-\gamma(1-m)+2} \to \infty$$

as $r \to -\infty$. Now, we are interested in knowing what happens as $\xi, r \to \infty$ to know what is the behaviour of the solutions when $|x| \to \infty$ and t is fixed and positive.

3.2.3 A further scaling property

Let $m \neq 1$. It is easy to check that whenever $f(\xi)$ is a solution of ODE (3.7), then for every $\lambda > 0$

$$\widehat{f}(\xi) = \lambda^{\mu} f(\lambda\xi) \tag{3.10}$$

is also a solution of the same equation if $\mu = 2/(1-m)$. This allows us to get a whole one-parameter family of solutions out of one of them. Thus, is $f(\xi) \sim \xi^{-\gamma}$ as either $\xi \to 0$ or $\xi \to \infty$, we get

$$\widehat{f}(\xi) \sim A\xi^{-\gamma} \qquad \text{with } A = \lambda^{\mu-\gamma}.$$

If $\gamma \neq 2/(1-m)$ we can adjust the coefficient A by means of λ. Let us see the effect of the scaling on the orbits. It is again easy to see that

$$\widehat{X}(\xi) = X(\lambda\xi), \quad \widehat{Y}(\xi) = Y(\lambda\xi),$$

hence we are travelling along the same orbit with a different parametrization (a scaling of ξ). In terms of $r = \log(\xi)$ it is even easier: we are just performing a translation of the parameter, the most elementary group invariance of an autonomous system. Note: this result is used in Theorems 3.2 and 5.11

3.2.4 Some special solutions: straight lines in phase plane

When looking for special solutions of the form (3.5), it is interesting to consider the solutions that correspond to straight line orbits in the (X, Y) plane. We have the trivial solutions $f = 0$, which correspond to orbits in the axis $Y = 0$. Apart from them we have

(i) The constant solutions $f = C$ occur for $\gamma = 0$ with $X = 0$, $Y > 0$.

(ii) Another vertical line solutions, $\dot{X} = 0$, which occur for $\gamma = (2-n)/m$, $X = -\gamma$ and variable $Y > 0$, so that

$$f^m = C\,|\xi|^{2-n}.$$

(iii) In the class of slanted lines the only admissible one occurs for $\gamma = n$ and takes the form

$$nX + \alpha Y = 0, \tag{3.11}$$

which for $\alpha > 0$ (which is equivalent to $m > m_c$) gives the ZKB solution. For completeness, let us say that there is also the possibility $\alpha < 0$ (equivalent to $m < m_c$), which gives a solution in the quadrant $Q = \{X > 0,\ Y > 0\}$ which is not globally defined in space. Its precise profile is

$$f^{1-m} = \frac{1}{(C - c|\xi|^2)_+},$$

where $C > 0$ is arbitrary and $c = |\alpha|(1-m)/2n$.

3.3 New special solution. Marcinkiewicz spaces

In this section we want to use the worst-case strategy in proving smoothing effect estimates for data in L^p spaces, $p > 1$. Now, it happens as before that the worst case is to be found outside of the L^p class. However, this time the worst case is given by the solution with scale-invariant initial data of the form

$$U_p(x,0) = \frac{C}{|x|^\gamma}, \qquad \gamma = \frac{n}{p}, \tag{3.12}$$

which lies in the Marcinkiewicz class $M^p(\mathbf{R}^n)$, and not in $L^p(\mathbf{R}^n)$.

Theorem 3.2 *There exists a unique solution of the PME with data* (3.12) *and it takes the self-similar form*

$$U_p(x,t) = t^{-\alpha_p} F(|x|\, t^{-\beta_p}) \tag{3.13}$$

where α_p is given in (3.2) *above, $\beta_p = p\alpha_p/n$. Moreover, the profile $F(\xi)$ is a bounded and rearranged function in $\mathcal{L}_0(\mathbf{R}^n)$. This profile behaves as $|\xi| \to \infty$ in the same way as the data* (3.12).

Proof The existence and uniqueness of a weak solution $u \in C([0,T) : L^1_{\rm loc}(\mathbf{R}^n))$ follows from well-known theory of the PME, see Section 1.2.1, since the data (3.12) belong to the class $\mathcal{X}_m \subset L^1_{\rm loc}(\mathbf{R}^n)$ for all $p > 1$. Using the equation scaling as explained in Section 3.2, we conclude that the solution must have the self-similar form

$$U(x,t) = t^{-\alpha} F(|x|\, t^{-\beta}). \tag{3.14}$$

In order to comply with the initial data $U_0(x) = A\,|x|^{-\gamma}$, the two parameters must satisfy the condition $\alpha = \beta\,\gamma$. Moreover, in order to satisfy the PME equation, the self-similar exponents must satisfy the relation

$$\alpha(m-1) + 2\beta = 1.$$

Both conditions imply the values

$$\beta = \frac{1}{2 + \gamma\,(m-1)}, \qquad \alpha = \frac{\gamma}{2 + \gamma\,(m-1)}, \tag{3.15}$$

which for our choice of $\gamma = n/p$ coincide with formulas (3.2). Note that $0 < \gamma < n$.

In order to conclude the proof we need some properties of the particular profile F. The important property for us here is boundedness: we need to prove that $F(0) = \sup\{F(\xi) : \xi \in \mathbf{R}^n\}$ is finite. This can be obtained in various ways.

(i) One way is applying the technique of phase-plane analysis. We reformulate the question of boundedness of F at $\xi = 0$ into an equivalent question about the orbit that represents F in the (X, Y) phase-plane analysis of the system. We have analysed that system in some detail in Section 3.2. Once the system is written in the form (3.9), the proof of existence of a bounded F is as follows: it is very easy to see that we can find an orbit in the (X, Y) phase plane starting from $(0, 0)$ and going into the second quarter of the plane, $Q = \{(X, Y) : \ X < 0, Y > 0\}$. We observe that it has a vertical asymptote at $X = -\gamma$. Translating the result into the properties of F as a function of ξ, we see that starting from $(0, 0)$ means that $F(\xi)$ is bounded at $\xi = 0$, and the asymptote means that the asymptotic behaviour is $F \sim C|\xi|^{-\gamma}$ as $\xi \to \infty$.

We refrain from going here into more details of this technique since it is a known result, and we will be more explicit later on in Section 5.4, where we will use a similar argument in order to show the existence of the self-similar solution on which we base the proof of the so-called backwards effect. Instead, we just display the numerical computation.

The interesting orbit we are talking about (Figure 3.1) goes from $(0, 0)$ to $(-\gamma, \infty)$ in the phase plane plot. We have taken

$$n = 3, m = 1/6, \gamma = 2.$$

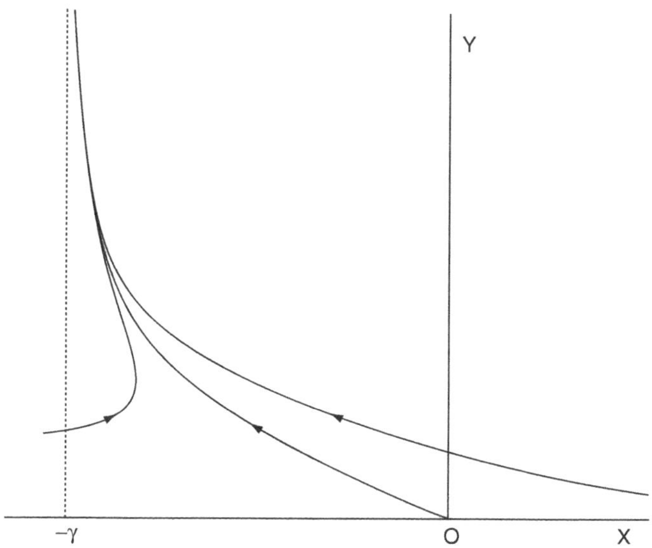

Figure 3.1 Phase-plane graphs of the self-similar solution.

They satisfy $n(1-m) > 2$ and $2/(1-m) > \gamma$. Then,

$$\alpha = \frac{\gamma}{\gamma(m-1)+2} = 6, \quad \beta = \frac{1}{\gamma(m-1)+2} = 3.$$

(ii) A second argument is as follows and will be preferred by those who like more standard ways of finding estimates in ODEs. We note that $F^m = G$ satisfies the ODE in $0 < r < \infty$:

$$(r^{n-1}G')' + r^{n-1}(\alpha F(r) + \beta r F'(r)) = 0, \quad F = G^{1/m}. \tag{3.16}$$

Elimination of the singularity at $r = 0$ can be proved by integral estimates under some integrability assumptions.

Lemma 3.3 *A solution $G(r) \geq 0$ of equation* (3.16) *defined for $0 < r < R$ is bounded when we assume that $G \in L^s(B_1(0))$ with $s > n(1-m)/2m$.*

These are precisely the assumptions that we are making. The proof of the estimate is easy by integration by parts in the spirit of Moser's iteration. It can be shown that the exponent in the assumption is optimal. □

3.4 New smoothing effect

Using the previous construction, we obtain the desired improvement of Theorem 3.1.

Theorem 3.4 *For every $u_0 \in M^p(\mathbf{R}^n)$, $1 < p < \infty$, and every $t > 0$ we have $u(t) \in L^\infty(\mathbf{R}^n)$ and*

$$|u(x,t)| \leq c(m,n,p,\infty)\, \|u_0\|_{M^p}^{\sigma_p}\, t^{-\alpha_p} \tag{3.17}$$

with α_p and σ_p given by (3.2). *The constant is attained at all times by the self-similar solution with data* (3.12).

We ask the reader to check the details of the proof using the line of argument of Theorem 2.1 and the fact that solution (3.13) has the most concentrated initial data among data with the same Marcinkiewicz norm.

Remark The best constant c in (3.1) need not be the same as the best constant we have identified in (3.17). In principle, it could be smaller.

We also the have the following consequence.

Theorem 3.5 *The map $S_t : u_0 \mapsto u(t)$ associated to the equation with $m > m_c$ is bounded from $M^p(\mathbf{R}^n)$ into itself. The constant of this map is 1. In other words, the PME/HE/FDE generates a bounded semigroup in $M^p(\mathbf{R}^n)$ for all $p > 1$ ($m > m_c$).*

Proof We only need to look at the self-similar solution. An easy calculation shows that this solution has constant M^p-norm. And the sup in the norm is taken as $x \to \infty$ (hint: use the fact that $Y = \xi^2 f^{1-m}$ is increasing along the orbit). The result follows. □

Remark The property of constant M^p-norm is another confirmation that these self-similar solutions play the role of the ZKB solutions in the context $p > 1$.

3.5 General smoothing result

We may now write the formula for the strong smoothing effects with correct exponents and best constants for all the transmission maps $u_0 \mapsto u(t)$ from $L^p(\mathbf{R}^n)$ to $L^q(\mathbf{R}^n)$ when $1 \le p \le q \le \infty$. It takes the form

$$\|u(t)\|_q \le c(m, n, p, q)\, \|u_0\|_p^{\sigma(p,q)}\, t^{-\alpha(p,q)} \tag{3.18}$$

with

$$\alpha(p, q) = \frac{n(q - p)}{q\,(n(m - 1) + 2p)}, \qquad \sigma(p, q) = \frac{p\,(n(m - 1) + 2q)}{q\,(n(m - 1) + 2p)}. \tag{3.19}$$

We leave it as an exercise for the reader to prove this general formula by interpolation of the previous cases, or if we want the best constant, by arguing that the worst case corresponds to the Marcinkiewicz function U_p whose behaviour in time is known through formula (3.13). □

We are now in a position to draw a general conclusion.

Theorem 3.6 *Assume that the data are taken in $L^p(\mathbf{R}^n)$ and the solution estimated in $L^q(\mathbf{R}^n)$ with $1 \le p, q \le \infty$. Then, a strong smoothing effect with time decay as $t \to \infty$ occurs iff $q \ge p$, and it has a decay of power type if $q > p$. If $p = 1$ the initial space can be extended to a measure space, while if $1 < p < q$ it can be extended to a Lorentz space. In all cases, $1 < q < \infty$ the end space is still a Lebesgue space.*

Proof (i) The positive part has already been proved for $q > p$. The case $p = q$ admits a smoothing effect with exponents $\alpha(p, p) = 0$ and $\sigma(p, p) = 1$, and constant $c \le 1$, since this means

$$\|u(t)\|_p \le \|u_0\|_p, \tag{3.20}$$

which is well-known monotonicity formula for the solutions of the PME (1.24).

(ii) For the negative result when $q < p$ we can give a simple proof when $p = \infty$, since constant data $u_0(x) = c$ produce constant solutions $u = c$ which are not in $L^q(\mathbf{R}^n)$, $q < \infty$ if $c > 0$.

(iii) For $p < \infty$ the negative argument is simple if we admit data in the Marcinkiewicz space $M^p(\mathbf{R}^n)$, $1 < p < \infty$, since then we may just consider the self-similar solutions U_p with data $U_p(x, 0) = |x|^{-n/p}$ described in Section 3.3. We know that for all times $t \ge 0$ this solution still belongs to $M^p(\mathbf{R}^n)$ and is moreover bounded, but it never belongs to $L^q(\mathbf{R}^n)$ with $q < p$ because of the precise behaviour as $|x| \to \infty$, $U_p(x, t) \sim |x|^{-\gamma}$ (same as the initial data). In fact, $U_p(\cdot, t)$ does not belong to $L^p(\mathbf{R}^n)$ either. □

Actually, modifying this example we get a stronger negative result for the possible smoothing effect going in the backward direction that we describe next.

Corollary 3.7 *Given $1 \le q < p \le \infty$, there exists a solution u with initial data $u_0 \in L^p(\mathbf{R}^n)$ such that for all $t > 0$ $u(\cdot, t) \notin M^q(\mathbf{R}^n)$. As a consequence, there can be no weak smoothing effect from $L^p(\mathbf{R}^n)$ into $M^q(\mathbf{R}^n)$ if $q < p$.*

Proof Let r be a value between q and p and let U_r be the solution with initial data $u_0(x) = |x|^{-n/r}$. We consider the solution

$$u(x,t) = U_r(x, t+\tau).$$

In view of the bounded behaviour of the solution and the behaviour at infinity, we conclude that $u(\cdot, t) \in L^p(\mathbf{R}^n)$ for all $p > r$ and all $t \geq 0$, but we also have

$$u(\cdot, t) \notin M^q(\mathbf{R}^n)\ \forall q < r, \forall t > 0.$$

This ends the proof and shows the power of self-similarity. □

Remarks

(1) Continuing with the argument on backward smoothing effects, it is worth noting the argument based on scaling that says that in a strong effect with $q < p$ the L^q norm must grow with time, and this goes against the monotone decreasing property of the solutions in all L^p norms. This is only an intuitive argument that we do not need to develop.

(2) The best constant $c(p,p) = 1$ even if data are taken in $L^p(\mathbf{R}^n)$. We leave it to the reader to check this fact.

(3) We will also skip the proof that the best constant depends continuously on the parameters m, n, p, and q when $q \geq p$.

3.6 The L^p–L^p problem. Estimates of weak type

Let us examine more closely the map $u_0 \mapsto u(t)$ from $L^p(\mathbf{R}^n)$ into itself. We know that

$$\|u(t)\|_p \leq \|u_0\|_p,$$

and this estimate is not improvable as a strong smoothing effect and corresponds to the best constant $c(p,p) = 1$. However, when a particular solution is considered then the estimate can be improved.

Theorem 3.8 *Let u be a solution of the PME with initial data $u_0 \in L^p(\mathbf{R}^n)$, $1 < p < \infty$. Then,*

$$\lim_{t\to\infty} \|u(t)\|_p \to 0. \tag{3.21}$$

Proof (i) By using the symmetrization and concentration results, it is sufficient to prove the result when $u_0 = u_0(r)$ is radially symmetric and rearranged. We can even assume that u_0 is bounded above in the form

$$u_0(r) \leq C\, r^{-n/p} \quad \text{for } r > 0, \tag{3.22}$$

and $u_0(r) r^{n/p} \to 0$ as $r \to \infty$. Under these assumptions we know that $u(x,t)$ is bounded for all times $t > 0$; actually, it is bounded above by the self-similar solution constructed in Theorem 3.2. We may also assume that u_0 is positive and smooth away from zero so that the solution is classical for all $t > 0$.

(ii) We consider the rescaled solutions

$$u_k(x,t) = k^{n/p} u(kx, k^{\lambda} t),$$

where, according to the scaling law (1.22) – (1.23), $\lambda = (m-1)(n/p)+2$. This scaling preserves the equation and the L^p norm of the solutions. We see that, as $k \to \infty$, the L^p norm is concentrated in a very small region around the origin so that $u_k(x,0)$ converges to zero in $L^p(\Omega_R)$, where Ω_R is the complement of any fixed ball $B_R(0)$, $R > 0$. Now, using Hölder's inequality we can check that $u_k(x,0)$ is also bounded in $L^1(B_R)$ for any fixed ball with a uniform estimate of the form $O(R^{n(p-1)/p})$. We conclude that $u_k(x,0)$ tends to zero in $L^1_{loc}(\mathbf{R}^n)$. Therefore, by the well-known continuity of solutions with respect to their data, the limit as $k \to \infty$ must be the trivial state $u = 0$ for all fixed $t > 0$, say, $t = 1$ (cf. Chapter 12 of [V06]).

(iii) In order to prove that the L^p norm of $u_k(\cdot, 1)$ goes to zero as $k \to \infty$ we only need to rule out a remaining possibility, i. e., that the bulk of the function moves to infinity. This is shown by controlling in a uniform way the L^p norm of u_k outside the ball of large but fixed radius R and proving that it goes to zero as $k \to \infty$. Indeed, this is true at $t = 0$ as a consequence of the form of the scaling. Next, multiplying the equation by a smooth cutoff function $\varphi(x) \ge 0$ that vanishes for $|x| \le R$ and is 1 for $|x| \ge R+1$ and integrating in space and time we get

$$\int u_k^p(x,t)\varphi(x)\,dx - \int u_k^p(x,0)\varphi(x)\,dx \le \int_0^1 \int p u^{m+p-2} \nabla u \cdot \nabla\varphi \,dxdt.$$

Integrating again by parts, using the fact that $|\Delta\varphi|$ is bounded and supported in $\{R \le |x| \le R+1\}$ and using the information on the initial data, we get

$$\int_{|x|\ge R} u_k^p(x,1)\varphi(x)\,dx \le \varepsilon_k + C_1 \int_0^1 \int_{R\le|x|\le R+1} u^{m+p-1}\,dxdt.$$

Taking into account that our solution is uniformly small at infinity because of estimate (3.22), we conclude that the last integral is small enough if R is large.

(iv) Undoing the scaling, the result follows as $t \to \infty$. □

Remark. This result is false for $p = 1$ or $p = \infty$.

The result implies an improvement of weak type for all $q > p$:

Corollary 3.9 *Let u be a solution of the PME with initial data $u_0 \in L^p(\mathbf{R}^n)$, $1 < p < \infty$. Then, for every $q > p$ we have*

$$\lim_{t\to\infty} \|u(t)\|_q \, t^{\alpha(p,q)} \to 0. \tag{3.23}$$

Therefore, the Marcinkiewicz functions U_p of formula (3.12) *are not asymptotic attractors for the solutions with data $u_0 \in L^p(\mathbf{R}^n)$.*

This is proved by using a two-step approach: we use the weak result of Theorem 3.8 from 0 to $t/2$ and then the standard result Theorem 3.6 from $t/2$ to t. We leave the easy details to the reader.

3.7 Negative results for $L^p(R^n)$, $0 < p < 1$

In our quest for generality of the scope of application of the preceding ideas, we might wonder if there is a smoothing effect for functions in the spaces $L^p(\mathbf{R}^n)$ with $0 < p < 1$, The answer is negative both for the strong and the weak effect when $m \ge 1$.

Theorem 3.10 *Let $u_0(x) \ge 0$ be a measurable function with compact support that is not locally integrable near a point $x_0 \in \mathbf{R}^n$. Then, there is no solution of the PME with such data. Moreover, if we approximate u_0 with integrable functions u_{0n} from below, the limit of the solutions $u_n(x,t)$ blows up instantaneously in the sense that*

$$\lim_{n\to\infty} u_n(x,t) = \infty \tag{3.24}$$

for all $x \in \mathbf{R}^n$ and $t > 0$. The same conclusion applies for the HE. In the case of the FDE in the range $m_c < m < 1$ the solutions exist but they are not bounded, not even locally integrable for $t > 0$.

Before proceeding with the proof, note that when $u_0 \ge 0$ is a measurable, compactly supported function, is locally bounded on balls away from $x = 0$ and near $x = 0$ behaves like $|x|^{-\gamma}$ with some $\gamma \ge n$, then $u_0 \in L^p(\mathbf{R}^n)$ for all $p < n/\gamma$, but it does not belong to $L^1_{\text{loc}}(\mathbf{R}^n)$, hence the solution does not exist and the conclusion of the Theorem applies.

Proof (i) The non-existence proof for the HE can be done by explicit computation of the integral representation of the solution, that we leave to the reader.

(ii) The non-existence result for $m > 1$ is a consequence of the Aronson–Caffarelli a priori estimate of Harnack type that says that for all non-negative solutions of the PME the value of u at a point x_0 and time t_0 is controlled from below by an explicit lower bound in terms of $\int_{B_r(x_0)} u(x,0)\,dx$, r, t_0. This weak Harnack inequality is explicitly stated and derived in Theorem 4.1 and we see that the lower bound for $u(x_0,t)$ diverges as the initial mass integral diverges.

(iii) The situation is different in the case of the FDE with $m > m_c$ since non-negative solutions with data in any of the spaces $L^p(\mathbf{R}^n)$ exist in the *extended continuous sense* introduced by Chasseigne and Vázquez in [ChV02]. However, such solutions are not locally bounded for $t > 0$ if $u_0 \notin L^1_{\text{loc}}(\mathbf{R}^n)$. Indeed, the local mass estimate of Herrero and Pierre [HP85] (see next section) says that whenever the $\int_{B_R(x_0)} u_0(x)\,dx = \infty$, then for every $t > 0$ the same divergence happens in any larger ball, which means that $u(t)$ cannot be locally bounded or belong to any $L^q_{\text{loc}}(\mathbf{R}^n)$ with $q \ge 1$. □

Remarks Dimensional considerations allow in principle for a formula of the form (3.1) with exponents given by (3.2), at least as long as $n(m-1)+2p > 0$. The conclusion is that the constant $c(m,n,p,\infty)$ must be infinite.

3.8 The question of local estimates for the FDE

The theory of the preceding chapters covers the main issues we had in mind for $m > m_c$, but there are still interesting aspects to consider. One of them concerns the existence of upper estimates in terms of local norms. The question whether it is possible to control $u(t)$ in $L^q_{\rm loc}(\mathbf{R}^n)$ when u_0 is controlled in $L^p_{\rm loc}(\mathbf{R}^n)$.

The answer to this problem (at least as stated) is negative for $m \geq 1$ as a consequence of Theorem 4.1 that says that you cannot get rid of the influence of masses located far away from your control domain. The same happens in the heat equation case for even more obvious reasons (superposition principle due to the linearity).

However, the local effects have been shown to be true for fast diffusion. Actually, they are one of the peculiar features of the fast diffusion range. This is due to the fact, already mentioned in Section 2.5 that the FDE is actually a 'slow diffusion equation' for large values of u.

Singular solution as barrier Our starting point is the observation that the definition of ZKB in the fast diffusion range can be extended to include the possibility of taking negative values for the free C in formula (2.8), i.e., considering the formula

$$U_m(x,t;M)^{1-m} = \frac{t}{k_1\,x^2 - C_1\,t^{2\alpha/n}}, \qquad (3.25)$$

with $\alpha = n/(2-n(1-m)) > 0$ and $k_1 = (1-m)\alpha/2n$ and $C_1 > 0$. The reader will easily check that this function defines a classical positive solution of the FDE in the range $m_c < m < 1$ in the external region

$$|x| > R(t) = (C_1/k_1)^{1/2}\,t^{\alpha/n}$$

which leaves inside an expanding hole of radius $R(t)$. Since this solution is infinite on the border of the hole, it serves as a universal barrier for the influence of any solution with data supported in the exterior of the hole.

Local estimates We begin by the celebrated local estimate for the FDE in the range $0 < m < 1$ obtained by Herrero and Pierre in [HP85] Lemma 3.1 that says that the L^1-norm of any solution at time $t > 0$ in a ball $B_R(a)$ is controlled in terms of the L^1-norm of u_0 in $B_{2R}(a)$, and conversely.

Theorem 3.11 *Let $m \in (0,1)$. There exists a constant $c = c(n,m) > 0$ such that for every weak solution $u(x,t) \geq 0$ of the FDE in $Q = \mathbf{R}^n \times (0,T)$ and every $0 < t, s < T$, $x_0 \in \mathbf{R}^n$ and $R > 0$ we have*

$$\int_{\{|x-x_0|\leq R\}} u(x,s)\,dx \leq c \int_{\{|x-x_0|\leq 2R\}} u(x,t)\,dx + c\,|t-s|^{\mu}\,R^{\sigma}, \qquad (3.26)$$

with $\mu = 1/(1-m)$, $\sigma = n - (2/(1-m))$. Note that $\sigma > 0$ if and only if $m < m_c$.

As a consequence, a solution must be locally integrable at $t > 0$ if and only if it already was at $t = 0$. A second estimate of that paper is the local smoothing effect L^1_{loc}-L^∞_{loc} that reads

Theorem 3.12 *For every $m_c < m < 1$ and every non-negative weak solution of the Cauchy problem u we have the estimate*

$$u(x,t) \leqslant C(m,n)\left[t^{-\theta}\left(\int_{B_r(x)} u_0(y)\,dy\right)^{2\theta/n} + (t/r^2)^{1/(1-m)}\right], \tag{3.27}$$

with $\theta = (m-1+(2/n))^{-1}$, $x \in \mathbb{R}^n$ and $t > 0$.

The local estimates are extended to the maximal solutions in the range $-1 < m \le 0$ for $n = 1$ in [ERV88]. We will discuss this issue in Section 8.1. Moreover, the L^∞ estimate does not hold for $m \le m_c$ as we will see.

The next question concerns local smoothing effects from L^p into L^∞. The result about necessary and sufficient conditions of boundedness is proved in [BU90] for all exponents $m < 1$, even $m \le 0$. In that paper the problem is addressed in terms of the so-called pressure variable, $v = u^{m-1}$, so that questions about boundedness of u translate into questions about positivity of v. The topic is investigated in detail in a local setting, see their Theorem 3.2 and Lemma 3.2.

We will give some hints about local estimates below, see for instance next chapter, but the whole issue deserves a separate study.

3.9 Comments, open problems, and notes

Most of the results of this chapter are new, very specially the use of Marcinkiewicz spaces as the natural setting for the smoothing estimates, Sections 3.4 and 3.5.

Section 3.1 This is a rather easy interpolation result, based on the preceding chapter. The L^p–L^∞ result for $p > 1$ is proved by DiBenedetto and Kwong [DK92], using different methods.

Section 3.3 The use of a general existence theorem with the scale invariance of the data to obtain existence of a self-similar solution, as it is done in Theorem 3.2, is a very beautiful and powerful argument. Such techniques are becoming now standard in works on the area (of nonlinear diffusion), but also in other areas of mechanics of continuous media.

Section 3.5 We will discover in Chapter 5 that backward effects are possible when the algebra of the similarity exponents is favourable, i.e., for $m < m_c$ and for p below the so-called critical line.

Section 3.6 The weak effects pose some interesting questions like the following.

Open problem 3.1 The exact decay rate for the solutions of Theorem 3.8 and Corollary 3.9 depends on the form of decay of the data at infinity, and there is no unique rate. In particular, if $u_0 \in L^1 \cap L^p$ the exact rate is known, cf. Theorem 2.1, and this is the fastest possible rate. In general the result is open. Question: is there a worst rate? Our guess is no.

Section 3.7 Further information on effects from and into L^p, $p < 1$:

For $p < 1$ both $L^p(\mathbf{R}^n)$ and $M^p(\mathbf{R}^n)$ are not good symmetrization spaces, so the technique we have been using fails.

Semi-explicit self-similar solutions give us the general picture; they have been examined in detail in [ChV02]. The behaviour depends on the so-called critical line $p_* = (1-m)n/2$ (this line will play a key role in the study of extinction for $m < m_c$).

Thus, for $p_* < p < 1$ any solution has a minimum behaviour as $|x| \to \infty$ of the form $u \geq c|x|^{-2/(1-m)}$ that makes the solutions belong in the best of cases to M^{p_*}, but not to M^p for any $p > p_*$. Lower estimates will be surveyed in the next chapter. When $0 < p < p_*$, the self-similar solutions with initial data $u_0(x) \sim C|x|^{-n/p}$ keep their behaviour in time, so no improvement of the M^p-space is possible.

More information on the issue of admissible initial data for the PME and related equations can be found in that paper.

Open problem 3.2 Does the PME flow map M^{p_*} into itself? For the separate-variables solution this is true and the quasi-norm is increasing in time (Hint: find the explicit solution).

Section 3.8 Using the two local estimates, a complete theory of singular solutions in the range $m_c < m < 1$ has been built in [ChV02]; we call such solutions *extended continuous solutions*; they can be obtained as limit solutions of standard smooth solutions, and have permanent singularities.

Exercises

Exercise 3.1 Prove that the functions U_p are asymptotic attractors of the solutions with data $u_0 \in L^p_{\text{loc}}(\mathbf{R}^n)$ that have the same behaviour at infinity as U_p.

Hint: use scaling, compactness and uniqueness of the limit.

Exercise 3.2 Examine the linear case $m = 1$, $p = 2$ for actual decay.

Hint: use Fourier analysis.

Exercise 3.3 Construct self-similar solutions to show that in the FDE case there is no improvement of the form L^p–L^q with $0 < p < q < 1$.

Exercise 3.4 Make precise the use of the universal barrier, formula (3.25) by proving the following result:

Let u_0 be an integrable function supported in $B_r(0)$, and let $u(x,t)$ be the solution of the FDE in the range $m \in (m_c, 1)$. Then, there exist constants t_0, A, $B > 0$ depending only on m, n and r, such that for every $|x| > r$ and every $t > 0$ we have

$$|u(x,t)|^{1-m} \leq \frac{t+t_0}{Ax^2 + B(t+t_0)^{2\alpha/n}}.$$

Calculate such constants.

4

Lower bounds, contractivity, error estimates, and continuity

We consider here a number of bounds that complement the study of previous chapters, which was centred on upper bounds.

The first section concerns the existence of lower bounds for non-negative solutions. There is a big difference between the ranges $m > 1$, $m = 1$, and $m \le 1$. In the first case, the property of finite propagation (cf. [Ar86]) implies that solutions may travel at a bounded speed so that it will take a certain time for a solution to become positive at points where it was initially zero. This does not happen for $m \le 1$ as we show.

The second section extends the property of contractivity in the L^1 norm into error estimates in different norms.

4.1 Lower bounds and Harnack inequalities

We obtain weak Harnack inequalities for our equations, in the two ranges, $m > 1$, and $m_c < m < 1$, starting from the well-known property valid for the linear heat equation, $m = 1$. We recall that the *classical Harnack inequality* states that any positive harmonic function defined in a ball $B_{2r}(0)$ of $\mathbf{R}^n$ satisfies the following inequality

$$\sup_{B_r(0)} u(x) \le C \inf_{B_r(0)} u(x),$$

where $C > 0$ is a constant that depends only on the dimension n. The result extends to non-negative solutions of the heat equation defined in a cylinder $Q = B_{2r}(0) \times (0, 4r^2)$ in the form

$$\sup_{Q_1} u(x,t) \le C \inf_{Q_2} u(x),$$

where again $C = C(n)$ and $Q_1 = B_r(0) \times (r^2, 2r^2)$, $Q_2 = B_r(0) \times (3r^2, 4r^2)$. Note that the cylinder Q_2 where the inf is taken comes later in time than the cylinder Q_1 where the sup is taken.

4.1.1 Estimating the eventual positivity for the PME

In the range $m > 1$ the equation has the property of finite propagation, cf. [Ar86], exemplified by the ZKB solutions that exhibit an infinite empty zone or zero-set $Z = \{(x,t) : u(x,t) = 0\}$ besides the occupied zone $\mathcal{P} = \{(x,t) : u(x,t) > 0\}$ (called positivity set), both separated by a sharp interface or free boundary $\Gamma = \partial\mathcal{P} \cap Q$. Therefore, the onset of positivity must take some time at points which lay at $t = 0$ inside the zero-set. This means that a Harnack inequality of the classical type cannot be true. There are versions that adapt very well to the properties of the nonlinear equations. To begin with, we remark that the ZKB solutions indicate that all points become eventually positivity points, and this property is proved for general non-negative solutions.[1]

The problem is then to find a quantitative statement of the eventual positivity of solutions. A natural inequality in this direction was obtained by Aronson and Caffarelli [AC83], and it has played a major role in developing the theory of the equation under general assumptions on the data.

Theorem 4.1 *There exists a constant $C = C(m, n) > 0$ such that the following estimate holds for all non-negative solutions of the PME:*

$$\int_{B_r(x_0)} u(x,0)\,dx \leqslant C\left(r^{1/\lambda(m-1)}T^{-1/(m-1)} + T^{n/2}u^{1/2\lambda}(x_0,T)\right) \tag{4.1}$$

with $\lambda = (n(m-1)+2)^{-1}$, for $x_0 \in \mathbf{R}^n$.

We present a version of the proof taken from [ChV02b] that uses in a direct way the smoothing effects we have derived, and is quite simple. We begin with a lemma when sizes are taken as unity.

Lemma 4.2 *Let $u \in C(\mathbb{R}^N \times [0,T])$ be a non-negative solution of the PME in $Q_1 = B_1(0) \times (0,1)$. Let*

$$M = \int_{B_1} u(x,0)\,dx. \tag{4.2}$$

There exist positive constants $M_0 = M_0(n,m)$ and $k = k(n,m)$ such that for $M \geqslant M_0$

$$u(0,1) \geqslant k\,M^{2\lambda}. \tag{4.3}$$

Proof This is a combination of several steps. The letter C will denote different positive constants that depend only on n and m.

(i) By comparison we may assume that u_0 is supported in the unit ball B_1. Indeed, a general u_0, is greater than $u_0\eta$, η being a suitable cut-off function compactly supported in B_1 and less than one. Thus, if v is the solution with initial data $u_0\eta$, we obtain

$$\int_{B_1} u(x,0)dx \geqslant \int_{B_1} u_0\eta = M,$$

and if the lemma holds true for v, then

$$u(0,1) \geqslant v(0,1) \geqslant kM^{2\lambda}.$$

We may then take the domain of definition as $Q = \mathbb{R}^n \times (0,\infty)$.

[1] See [V06], Chapter 14, for more details on the issue of finite propagation.

(ii) By using the standard smoothing effect, we know the a priori estimate for the solution

$$0 \leqslant u(x,t) \leqslant C\, M^{2\lambda}\, t^{-n\lambda}. \tag{4.4}$$

We also have the a priori estimate for the support at time t (cf. [CVW87] or [V06], Chapter 18)

$$\operatorname{supp} u(\cdot,t) \subset B_R(t), \qquad R(t) = C\, M^{(m-1)\lambda}\, t^{\lambda}. \tag{4.5}$$

Note that if M is large this radius is much larger than 1 at $t = 1$.

(iii) Aleksandrov's reflection argument, as used in Lemma 2.2 of [AC83], means that for $|x| \geq 2$ we have

$$u(0,t) \geqslant u(x,t). \tag{4.6}$$

(iv) By conservation of mass we know that

$$\int_{\mathbf{R}^n} u(x,t)\,dx = \int_{\mathbf{R}^n} u_0(x)\,dx. \tag{4.7}$$

The left-hand side can be split into the integrals

$$\int_{|x|\geqslant 2} u(x,t)\,dx + \int_{|x|\leqslant 2} u(x,t)\,dx,$$

where the last term can be estimated by

$$C\, M^{2\lambda}\, t^{-n\lambda} 2^n.$$

We conclude that

$$Cu(0,t)\,(R(t)^n - 2^n) \geqslant \int_{|x|\geq 2} u(x,t)\,dx \geqslant M - C\, M^{2\lambda}\, t^{-n\lambda} 2^n,$$

hence for $t = 1$,

$$C\, u(0,1)\,(M^{n(m-1)\lambda} - 2^n) \geqslant M - C\, M^{2\lambda}.$$

If $M > 1$ there are constants c_1, c_2 depending on (m,n) such that

$$u(0,1) \geqslant c_1 M^{2\lambda} - c_2 M^{\gamma\lambda}, \quad \gamma = 2 - (m-1)n.$$

Since $\gamma < 2$, there exist constants M_0 and k such that

$$c_1 M^{2\lambda} - c_2 M^{\gamma\lambda} \geqslant k M^{2\lambda}$$

holds for every $M \geqslant M_0$, and this proves the Lemma. □

We can now prove the full Harnack-type inequality:

Proof of Theorem 4.1 We perform the transformation

$$u^*(x,t) = r^{-2/(m-1)} T^{1/(m-1)} u(rx, Tt). \tag{4.8}$$

We easily see that u^* satisfies the same equation, so can use the previous lemma on u^*. Noting that

$$\int_{|x|\le 1} u^*(y,t)dy = r^{-(n+\frac{2}{m-1})}T^{\frac{1}{m-1}} \int_{|x|\le r} u(x,Tt)\,dx,$$

and applying the already derived formula (4.3) to u^*, we conclude that

$$\int_{B_r(0)} u(x,0)\,dx \leqslant C\,T^{n/2}u^{1/2\lambda}(0,T)$$

on the condition that

$$\int_{B_r(0)} u(x,0)\,dx \ge r^{1/\lambda(m-1)}T^{-1/(m-1)}M_0.$$

Formula (4.1) is equivalent to the final conclusion for $x_0 = 0$. But the choice of origin is indifferent. □

Indeed, there is a way of writing the result that is more interesting for exhibiting the positivity property. Let u be a solution with initial data $u_0 \ge 0$ and let

$$M_R(x_0) = \int_{B_R(x_0)} u_0(x)\,dx.$$

Then we have the following reformulation of Theorem 4.1.

Theorem 4.3 *For every solution with locally integrable data and for every $R, t > 0$ and $x_0 \in \mathbf{R}^n$ there are constants $c_1, c_2 > 0$ depending only on m, n such that*

$$u(x_0,t) \ge c_1\,M_R(x_0)^{2\lambda}t^{-n\lambda} \tag{4.9}$$

if the time is not too small, precisely when

$$t \ge t_* = c_2 R^{1/\lambda} M_R(x_0)^{1-m}. \tag{4.10}$$

Remarks

(1) Since $\alpha = n\lambda$ is the exponent in the L^1–L^∞ smoothing effect, we have found that for large times an estimate similar to the upper estimate of the L^1–L^∞ is proved. However, the time where it begins to apply at a point x_0 depends on the initial mass around x_0 in an inverse power way.

(2) The exponents in the expressions are exact, since they have been obtained by scaling considerations. However, we do not give a method to calculate the best constants. R plays the role of a free parameter.

(3) The expression for the *characteristic time* t_* of the positivity effect is related to the *waiting time*, i.e., to the time a solution that starts with zero initial data in a ball $B_R(x_0)$ spends until $u(x_0,t)$ becomes positive at x_0

Corollary 4.4 *In particular when*

$$\lim \frac{M_R(x_0)}{R^{n+2/(m-1)}} = \infty \tag{4.11}$$

as $R \to 0$ or $R \to \infty$, then $u(x_0, t)$ is positive for all $t > 0$.

This condition of zero waiting time has been proved to be sharp in [CVW87].

(4) We could think about a better expression (more optimized) for the lower bound of the form

$$u(x_0, t) \geq H(t, M_R(x_0), R), \tag{4.12}$$

valid for all $t > 0$, and H being a continuous function. Scaling considerations imply then that the formula takes the form

$$u(x_0, t) \geq \frac{M_R(x_0)}{R^n} H(t/t_*) = \overline{M}_R(x_0)\, H(t/t_*), \tag{4.13}$$

where $\overline{M}_R(x_0) = ⨏_{B_{R(x_0)}} u_0(x)\, dx = \int_{B_{R(x_0)}} u_0(x)\, dx / R^n$. We refer to Exercise 4.1 for a proof of this reduction. Finding an optimal formula is an open problem. Note that our estimate above gives a broken curve.

4.1.2 The heat equation case

An optimal lower estimate of the type we are discussing is very easily obtained in the case of the heat equation. It says

Proposition 4.5 *If u is a non-negative solution of the HE in $Q = \mathbf{R}^n \times (0, T)$ then for all $t, R > 0$ and $x_0 \in \mathbf{R}^n$ we have*

$$u(x_0, t) \geq (4\pi)^{-n/2} M_R(x_0)\, t^{-n/2}\, e^{-R^2/4t} \tag{4.14}$$

and the exponents and constant are optimal.

Proof The representation formula says that

$$u(x_0, t) \geq (4\pi t)^{-n/2} \int_{\mathbf{R}^n} u_0(y)\, e^{-(y-x_0)^2/4t}\, dy,$$

so that the best estimate in terms of $M_R(x_0)$ consists in forgetting the part of u_0 not supported in $B_R(x_0)$ and displacing all the mass in this ball to the boundary. We get the solution corresponding to a M times the Dirac delta located at a point of the boundary. □

In comparison with the results of the previous subsection, we see that here that the characteristic time is $t_* = cR^2$ and that formula (4.13) is respected. We notice that this positivity estimate consists of two time periods: an increasing lower bound for an initial time interval $0 < t < t_c$ plus an exponentially decreasing bound for all later times. In

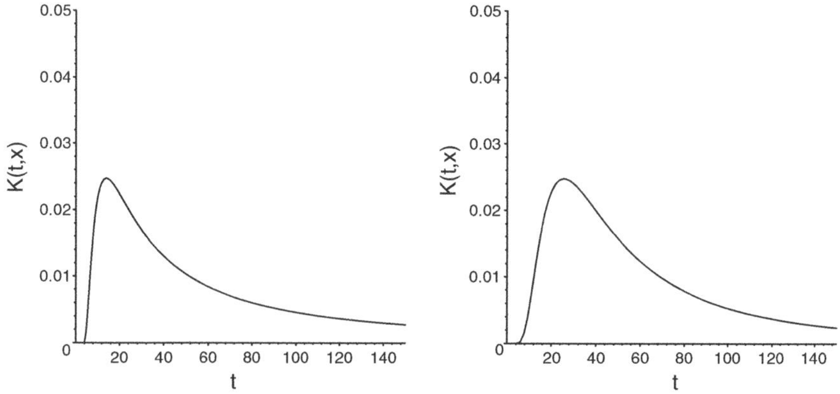

Figure 4.1 Lower bound functions $H(t)$ when $m > 1$ and $m = 1$.

the case $m > 1$ we cannot ensure positivity in the initial period where waiting times may occur.

There is a difference to be remarked in the form the profiles of both graphs start near $t = 0$ (see Figure 4.1). In the PME case there must be a clear waiting time, while in the heat equation case we have very slow growth but no waiting time.

4.1.3 Positivity estimates for fast diffusion when $m_c < m < 1$

We now derive positivity estimates for the fast diffusion equation in the upper range $(n-2)_+/2 = m_c < m < 1$. Recall that $\lambda = (2 - n(1-m))^{-1}$ is positive in this range. We will still get an estimate with two periods of different monotonicities and calculate everything with sharp exponents. Moreover, we will also show that the description in two periods reflects a real situation for some typical initial data. Actually, the explicit computation for the Barenblatt solution in the worst case where the initial mass is placed on the border of the ball B_{R_0} gives

$$U(0,t) = \frac{M_{R_0}^{2\lambda}\, t^{1/(1-m)}}{(c_1 t^{2\lambda} + c_2 t_c^{2\lambda})^{1/(1-m)}}, \tag{4.15}$$

which shows in a very clear way the two patterns: an increasing behaviour for an initial time interval followed by a decay, both with power rates that are reproduced by the following general result (Figure 4.2).

Theorem 4.6 *Let $u \in C(Q)$ be a non-negative solution to $u_t = \Delta(u^m)$ in Q, $Q = [0,+\infty) \times \mathbf{R}^n$, with $m_c < m < 1$ with initial data $u_0 \in L^1_{\mathrm{loc}}(\mathbf{R}^n)$, $u_0 \geq 0$. There exists a positive function $H(t)$ such that for any $t > 0$ and $R > 0$ the following bound holds true:*

$$\inf_{x \in B_R} u(x,t) \geq \overline{M}_R H(t/t_c), \tag{4.16}$$

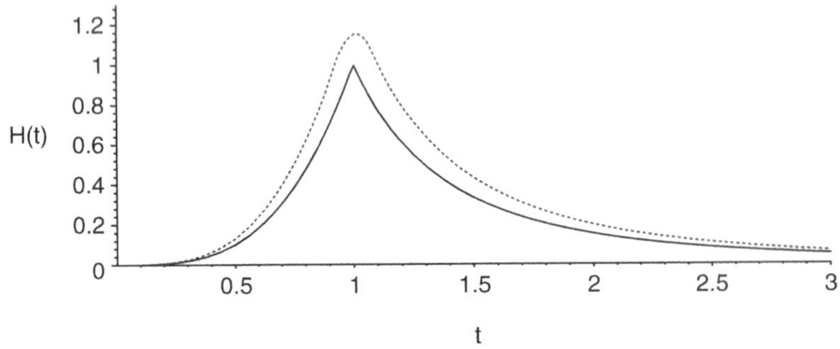

Figure 4.2 The function $H(t)$ when $m < 1$ and a solution profile.

where function $H(s)$ is positive and takes the precise form

$$H(s) = \begin{cases} K s^{1/(1-m)} & \textit{for } s \le 1, \\ K s^{-n\lambda} & \textit{for } s \ge 1. \end{cases} \tag{4.17}$$

The critical time is given again by formula $t_c = C\, M_R^{1-m} R^{1/\lambda}$. Constants $C, K > 0$ depend only on n and m.

Proof The proof is a combination of several steps. Different positive constants that depend on m and n are denoted by C_i. The precise value we get for the constants C and K in the statement is given at the end of the proof.

(i) *Reduction.* As in Lemma 4.2, we may assume $\mathrm{supp}(u_0) \subset B_{R_0}(0)$ for some radius R_0 by comparing u with the solution v of the same fast diffusion equation with initial data $u_0\eta$, η being a suitable cutoff function compactly supported in B_{R_0} and less than one. We obtain

$$\int_{B_{R_0}} u(x,t)\,dx \ge \int_{B_{R_0}} u_0(x)\eta(x)\,dx = M_{R_0}$$

and if the lemma holds true for v, then

$$\inf_{x \in B_{R_0}} u(x,t) \ge \inf_{x \in B_{R_0}} v(x,t) \ge H(t) M_{R_0}^{\lambda}.$$

(ii) *A priori estimates* The second step is based on the well-known smoothing estimates

$$u(x,t) \le C_1 \|u_0\|_1^{2/(2+n(m-1))} t^{-n/(2+n(m-1))}.$$

We remark that $\|u_0\|_1 = M_{R_0}$ since u is non-negative and supported in B_{R_0}, so that we get

$$u(x,t) \le C_1 M_{R_0}^{2\lambda} t^{-n\lambda}$$

for any $x \in \mathbf{R}^n$, with $\lambda = 1/(2 + n(m-1))$. Integration over B_{R_0} gives

$$\int_{B_{R_0}} u(x,t)\,dx \le C_2\, M_{R_0}^{2\lambda} R_0^n\, t^{-n\lambda} \tag{4.18}$$

where $C_2 = C_1\omega_n$.

(iii) *Integral estimate.* The third step uses a Herrero-Pierre L^1_{loc} estimate, cf. Theorem 3.11: for any $R > 0$ and $s, t \ge 0$ one has

$$\int_{B_{R_0}} u(x,s)\,dx \le C_3 \left[\int_{B_{R_0+R}} u(x,t)\,dx + \frac{|s-t|^{1/(1-m)}}{R^{(2-n(1-m))/(1-m)}} \right].$$

We let $s = 0$ and we rewrite it in a form more useful to our purposes:

$$\int_{B_{R_0+R}} u(x,t)\,dx \ge \frac{M_{R_0}}{C_3} - \frac{t^{\frac{1}{1-m}}}{R^{\frac{2-n(1-m)}{1-m}}}. \tag{4.19}$$

(iv) *Aleksandrov principle.* Arguing as for the case $m > 1$, we have $u(x_0,t) \ge u(x_1,t)$ if $|x_0| < R_0$ and $|x_1| > R_0$, whenever u is as in our hypothesis and $t > 0$. From this, it easily follows that:

$$\int_{B_{R_0+R}\setminus B_{R_0}} u(x,t)\,dx \le A_n R^n \inf_{x\in B_{R_0}} u(x,t), \tag{4.20}$$

where A_n is chosen so that $A_n R^n \ge |B_{R_0+R} \setminus B_{R_0}|$; we have $1 < A_n < 2^n$.

(v) We now put together all the previous calculations:

$$\begin{aligned}\int_{B_{R_0+R}} u(x,t)\,dx &= \int_{B_{R_0}} u(x,t)\,dx + \int_{B_{R_0+R}\setminus B_{R_0}} u(x,t)\,dx \\ &\le C_2 \frac{M_{R_0}^{2\lambda} R_0^n}{t^{n\lambda}} + A_n R^n \inf_{x\in B_{R_0}} u(x,t).\end{aligned}$$

This follows by (4.18) and (4.20). Using now (4.19), we obtain

$$\frac{M_{R_0}}{C_3} - \frac{t^{\frac{1}{1-m}}}{R^{\frac{2-n(1-m)}{1-m}}} \le \int_{B_{R_0+R}} u(x,t)\,dx \le C_2 \frac{M_{R_0}^{2\lambda} R_0^n}{t^{n\lambda}} + A_n R^n \inf_{x\in B_{R_0}} u(x,t),$$

and finally,

$$\inf_{x\in B_{R_0}} u(x,t) \ge \frac{1}{A_n}\left[\left(\frac{M_{R_0}}{C_3} - C_2 \frac{M_{R_0}^{2\lambda} R_0^n}{t^{n\lambda}} \right) \frac{1}{R^n} - \frac{t^{\frac{1}{1-m}}}{R^{\frac{2}{1-m}}} \right]. \tag{4.21}$$

(vi) Now we would like to obtain the claimed estimate for $t > t_c^*$. To this end we first check whether the grouping of the first two terms is positive:

$$B(t) := \frac{M_{R_0}}{C_3} - C_2 \frac{M_{R_0}^{2\lambda} R_0^n}{t^{d\lambda}} > 0 \Longleftrightarrow t > (C_3 C_2)^{1/(d\lambda)} M_{R_0}^{1-m} R_0^{1/\lambda} = t_c^*. \quad (4.22)$$

Assuming then that $t \geq t_c = 2t_c^*$ (so that $B(t_c)$ is clearly positive), we optimize the function

$$f(r) = \frac{1}{A_d} \left[\frac{B(t)}{r^n} - \frac{t^{\frac{1}{1-m}}}{r^{\frac{2}{1-m}}} \right]$$

with respect to $r > 0$ and we obtain that it attains its maximum at $r = r_{\max}$, and after a few straightforward computations we obtain that its maximum value is:

$$f(r_{\max}) = A_d \left[\frac{d(1-m)}{2} \right]^{2\lambda} \left[\frac{2}{d(1-m)} - 1 \right] \left[\frac{1}{C_3} - C_2 \frac{M_{R_0}^{2\lambda-1} R_0^n}{t^{d\lambda}} \right]^{2\lambda} \frac{M_{R_0}^{2\lambda}}{t^{d\lambda}} > 0.$$

We have thus found the estimate

$$\begin{aligned} \inf_{x \in B_{R_0}} u(x,t) &\geq A_d \left[\frac{d(1-m)}{2} \right]^{2\lambda} \left[\frac{2}{d(1-m)} - 1 \right] \left[\frac{1}{C_3} - C_2 \frac{M_{R_0}^{2\lambda-1} R_0^n}{t^{d\lambda}} \right]^{2\lambda} \frac{M_{R_0}^{2\lambda}}{t^{d\lambda}} \\ &= K_1 H_1(t) \frac{M_{R_0}^{2\lambda}}{t^{d\lambda}}. \end{aligned}$$

A straightforward calculation shows that the function

$$H_1(t) = \left[\frac{1}{C_3} - C_2 \frac{M_{R_0}^{2\lambda-1} R_0^n}{t^{d\lambda}} \right]^{2\lambda}$$

is non-decreasing in time, thus if $t \geq t_c$:

$$H_1(t) \geq H_1(t_c) = \left(\frac{1}{2C_3} \right)^{2\lambda}$$

and finally we obtain:

$$\inf_{x \in B_{R_0}} u(x,t) \geq K_1 H_1(t) \frac{M_{R_0}^{2\lambda}}{t^{d\lambda}} \geq K_1 H_1(t_c) \frac{M_{R_0}^{2\lambda}}{t^{d\lambda}} = \frac{K}{(2C_3)^{2\lambda}} \frac{M_{R_0}^{2\lambda}}{t^{d\lambda}}.$$

This is exactly (4.16) when $t > t_c$.

(vii) The last step consists in obtaining a lower estimate when $0 \le t \le t_c$. To this end we consider the fundamental Bénilan–Crandall estimate [BC81b]:

$$u_t(x,t) \le \frac{u(x,t)}{(1-m)t}$$

which easily implies that the function

$$u(x,t)t^{-1/(1-m)}$$

is non-increasing in time, thus for any $t \in (0, t_c)$ we have that

$$u(x,t) \ge \frac{u(t_c,x)}{t_c^{1/(1-m)}} t^{1/(1-m)}.$$

In order to obtain inequality (4.16) for $0 < t < t_c$ it is now sufficient to apply the inequality valid for $t > t_c$ to the right-hand side of the above inequality. Replacing R by R_0, the proof of formula (4.16) is complete in all cases. Constant C has the value $C = 2\,(C_3C_2)^{1/(n\lambda)}$ while K is given by

$$K = \frac{2^{-(n+2)\lambda}}{A_n\, C_2\, C_3^{2\lambda+1}} \left[\frac{n(1-m)}{2}\right]^{2\lambda} \left[\frac{2}{n(1-m)} - 1\right]. \tag{4.23}$$

□

The explicit computation for the Barenblatt solution done above shows that the behaviour of H is optimal in the limits $t \gg 1$ and $t \approx 0$.

Harnack inequality for large times In fact, joining the upper and lower estimates, we have a rather perfect inequality of this type for large times $t \ge t_c$:

$$C_1 M_{R_0}^{2\lambda}\, t^{n\lambda} \le u(x,t) \le C_2 M^{2\lambda}\, t^{n\lambda}, \tag{4.24}$$

where M is the total mass $M = \int_{\mathbf{R}^n} u_0(x)\,dx$. Moreover, $u(0,t)$ can be replaced by $u(x,t)$ with x in any bounded set (the constants will depend on the set).

A minimax problem Suppose that we want to obtain the best of the lower bounds when t varies with given R and M_R. This happens for $t/t_c \approx 1$ and the value is

$$u(0,t_1) \ge C_3\, M_R R^{-d},$$

which is just proportional to the average. At this time also the maximum is controlled by the average.

4.1.4 Harnack inequality for FDE on R^n

The local positivity result for the FDE can be refined in two ways; first of all, it can be combined with the already known upper bounds of the smoothing effect to obtain a full local Harnack inequality; on the other hand, it can be combined with the infinite speed of propagation to obtain a global Harnack principle on the whole space. We review here the results obtained in those directions by Bonforte and Vázquez in [BV05].

Theorem 4.7 (Elliptic Harnack inequality) *Let $u(t,x)$ satisfy the same hypothesis as Theorem 4.6. If moreover $u_0 \in L^1(\mathbf{R}^n)$, there exists a positive constant $\mathcal{H}$, depending only on m and d on the ratio M_R/M_∞, such that for any $t \geq t_c(M_R, R)$:*

$$\sup_{x \in B_R} u(x,t) \leq \mathcal{H} \inf_{x \in B_R} u(x,t). \tag{4.25}$$

If moreover u_0 is supported in B_R, then the constant $\mathcal{H}$ is universal and depends only on m and d.

We have put

$$M_\infty = \int_{\mathbf{R}^n} u_0(x)\,dx, \qquad M_R = \int_{B_R} u_0(x)\,dx \tag{4.26}$$

for some $R > 0$, $x_0 \in \mathbf{R}^n$. Under a further control on the initial data, we can transform the local Harnack principle into a global version. We recall that b_i, λ_1, k_1, and C_i are constants that depend only on m and d, while rest of the parameters depend also on the data as expressed.

Theorem 4.8 (Global Harnack principle)
Let $u_0 \in L^1(\mathbf{R}^n)$, $u_0 \geq 0$ and

$$u_0(x)\,|x|^{2/(1-m)} \leq A \tag{4.27}$$

for $|x| \geq R_0$. Then, for any time $\varepsilon > 0$ there exist constants τ_1, τ_2, M_1 and M_2, such that for any $(t,x) \in (\varepsilon, \infty) \times \mathbf{R}^n$ we have the following upper and lower bounds:

$$\mathcal{B}(x, t-\tau_1; M_1) \leq u(t,x) \leq \mathcal{B}(x, t+\tau_2; M_2), \tag{4.28}$$

where $\tau_1 = \lambda_1 \varepsilon$, $\tau_2 = \tau(\varepsilon, A, t_s)$, $M_1 = M(\varepsilon)$ and $M_2 = k_2(\varepsilon, A, \tau_2)\, M_\infty$, while

$$t_c = C\, M_R^{1-m}\, R^{1/\lambda}, \qquad t_s = C_5\, M_\infty^{1-m}\, R_0^{1/\lambda}.$$

We point out that b_i, λ_1, k_1, and C_i are constants that depend only on m and n, while the rest of the parameters depend also on the data as expressed. We refrain from entering here into the proofs of these results that can be consulted in the mentioned reference.

4.2 Contractivity and error estimates

As we have said in the Introduction, for linear equations any boundedness estimate is equivalent to a stability result, In that case, the improvement L^1 into L^∞ is called ultracontractivity and the lesser improvement L^2 into L^4 is called hypercontractivity [Dv89, Bk94]. Since equation $u_t = \Delta u^m$ is nonlinear for $m \neq 1$, the estimates that give boundedness of the PME/FDE flow do not imply as a direct consequence any kind of contractivity, since this one must be formulated in terms of differences.

A standard approach to contractivity in the PME/FDE theory comes from the well property that flow is contractive in the L^1 norm, as explained in Section 1.2, formula (1.25). It will be convenient for us to view contractivity properties as a form of error estimates, which in turn imply a quantitative formulation of the continuity of the PME/FDE semigroup. We are able to obtain a control of the L^q norm of the difference of two solutions by using Hölder's inequality,

$$\|u_1(t) - u_2(t)\|_q^q \le \|u_1(t) - u_2(t)\|_1 \|u_1(t) - u_2(t)\|_\infty^{q-1}, \tag{4.29}$$

plus the estimate of the L^1-L^∞ smoothing effect. This gives a first error estimate.

Lemma 4.9 *Let u_1 and u_2 be two solutions of equation* (2.1) *in the range $m > m_c$ with initial data $u_{01}, u_{02} \in L^1(\mathbf{R}^n)$. Then, for every $1 < q < \infty$ there is a constant $C(m, n, q) > 0$ such that*

$$\|u_1(t) - u_2(t)\|_q \le C(m, n, q)\, \|u_{01} - u_{02}\|_1^{1/q} \|u_{01} + u_{02}\|_1^{\sigma'} t^{-\alpha'}, \tag{4.30}$$

with $\alpha' = \alpha(q-1)/q$, $\sigma' = 2(q-1)\alpha/(nq)$, and α is the decay exponent given in (2.3). *Consequently, the PME/FDE semigroup is continuous from $L^1(\mathbf{R}^n)$ into $L^q(\mathbf{R}^n)$ for all $q < \infty$.*

Such a simple procedure does not allow us to get an error estimate in the L^∞ norm. Such an estimate depends on knowing more about the regularity of the solutions. Actually, we will obtain an improved error estimate by making use of the result about uniform Hölder continuity of bounded solutions stated in Section 1.2, formula (1.21): if $u(x, t)$ is solution defined in the strip $\{(x, t) : 1/2 \le t \le 1\}$ and bounded in it by 1, then

$$|u(x, 1) - u(x', 1)| \le C|x - x'|^\mu$$

for some constants $C > 0$ and $\mu \in (0, 1)$ that depend on m and n. An accurate estimate of the Hölder constant for a general solution $u(x, t)$ at any time $T > 0$ can be obtained by using famous scaling transformation (1.22)–(1.23) in the way it was used in Proposition 2.2. We need to assume a certain decay of the L^∞-norm in time, like

$$u(x, t') \le M(t) \quad \forall\, 0 < t < t'. \tag{4.31}$$

Then we have

Lemma 4.10 *Let u be a solution of the PME with $m > 0$ satisfying assumption* (4.31). *Then the Hölder constant at time t can be estimated above by*

$$C = C_h(m, n) M(t/2)^\nu t^{-\mu/2}, \quad \nu = \frac{2 - (m-1)\mu}{2}, \tag{4.32}$$

and the exponents are sharp.

Proof We fix $T > 0$ and take $\tilde{u}(x,t) = Ku(Lx, Tt)$ with $K = M(T/2)^{-1}$ so that

$$0 \le \tilde{u}(x,t) \le 1 \quad \text{for all } t > 1/2.$$

Then, the regularity result (1.21) applies. In terms of u, it means that

$$|u(x,T) - u(x',T)| \le \frac{C_h}{KL^{\mu}}|x - x'|^{\mu}.$$

Since we have $L^2 = K^{1-m}T$ by (1.23), the result follows replacing T by t. □

We are now ready to state and prove a sharper continuity result for the semigroup in the form of error estimate.

Theorem 4.11 *Let u_1 and u_2 be two solutions of equation (2.1) with initial data $u_{01}, u_{02} \in L^p(\mathbf{R}^n)$ for some $p \ge 1$ and $m > m_c$. Let $\|u_{01} - u_{02}\|_1$ be finite. Then, we have an error estimate of the form*

$$\|u_1(t) - u_2(t)\|_\infty \le C\, \|u_{01} - u_{02}\|_1^{\frac{\mu}{\mu+n}} N^{\lambda} t^{-\gamma}, \tag{4.33}$$

for some $\gamma, \lambda, C > 0$ depending on (m, n, p), and $N = \max\{\|u_{01}\|_p, \|u_{02}\|_p\}$.

Proof By the smoothing effect, the solutions are bounded with an estimate of the form

$$M(t) = C\, N^{\sigma_p} t^{-\alpha_p}, \tag{4.34}$$

in the notation of Theorem 3.4. It follows from the previous lemma that they are uniformly Hölder continuous and satisfy estimate (4.32). We also have a control of the L^1-error

$$\|u_1(t) - u_2(t)\|_1 \le \|u_{01} - u_{02}\|_1 = \epsilon.$$

We have to translate this into an estimate in L^∞ error using the Hölder continuity. This is done by means of a simple calculus lemma that goes as follows: if we put $\|u_1(t) - u_2(t)\|_\infty = \delta$, and $C(t)$ is the Hölder constant with exponent μ at time t, then there is a (space) region of size R where the error is at least $\delta/2$ between $u_1(\cdot, t)$ and $u_2(\cdot, t)$ if $C(t)R^\mu \le \delta/4$. In that region the L^1-error is easily estimated as $O(R^n\delta)$. Therefore, $R^n\delta \le c\epsilon$. Taking the value of R into account we get

$$\delta \le c\, C(t)^{\frac{n}{\mu+n}} \epsilon^{\frac{\mu}{\mu+n}},$$

and $c = c(m,n)$. Since we know the value of $C(t)$ from Lemma 4.10, and $M(t)$ is given by (4.34), we get

$$C(t) = C_h M(t/2)^{\nu} t^{-\mu/2} = C_2 N^{\sigma_p \nu} t^{-\alpha_p \nu - \mu/2}$$

and C_2 depends m and n. This implies the result. The precise values of γ and λ calculated here are

$$\gamma = \frac{n(n+\mu p)}{(\mu+n)(2p+n(m-1))}, \qquad \lambda = \frac{n\sigma_p \nu}{\mu+n}, \tag{4.35}$$

with ν given by (4.32). They need not be optimal. Note that $\mu/(n+\mu) \to 0$, $\nu \to 1$, and $\gamma \to \alpha_p$ as $\mu \to 0$, thus obtaining the exponents of formula (3.17), as expected. □

Remarks

(1) This result allows us to improve the previous L^q estimates using Hölder's inequality (4.29). We leave the details as a further exercise to the reader.

(2) The condition $m > m_c$ is only used to ensure the L^p–L^∞ effect. This effect holds also for m less than m_c for large p, see next chapter.

4.3 Comments and historical notes

Section 4.1 Harnack inequalities were first proved for harmonic functions and then for uniformly elliptic equations, the classical reference being Moser's [Mo61]. They were then extended to the heat equation and parabolic operators, cf. [Mo64] or [Tr68]. It is known that such a Harnack inequality is strongly related to detailed estimates of the heat kernel, i.e., the fundamental solution of the heat equation, as the ones obtained by Aronson [Ar67].

Subsection 4.1.1 The Harnack inequality for the PME we present is due to Aronson and Caffarelli [AC83] who used it to show the existence of initial traces for all non-negative solutions defined in a strip $Q = \mathbf{R}^n \times (0, T)$. Bénilan, Crandall and Pierre [BCP] proved a result on optimal conditions of existence that complements the estimate.

The proof of the Harnack inequality we present in Lemma 4.2 was written in [ChV02b] for the more general equation $u_t = \Delta u^m - au^p$, with $a \geq 0$ and $p \leq 1 < m$. The Aleksandrov reflection argument is well known: it is, for instance, explained in [GV03], pages 51–52.

The waiting time is an aspect of the positivity question that has been much studied in the PME literature because the situation of the free boundary is in those cases metastable (it is stationary for a time and then starts to move). The condition for showing zero waiting time given in Corollary 4.4 has been proved to be sharp in [CVW87]. A detailed study of waiting times is performed in [V06], Chapters 14 and 15.

Once we know that solutions are bounded and positive, the equation is no longer degenerate at the local level and a regularity theory may be developed to prove that solutions are C^∞ along the lines of standard quasilinear theory, cf. [F64, LSU]. But even in the general case where solutions are only assumed to be non-negative, it can be proved that solutions are indeed C^α continuous with a certain exponent $\alpha \in (0, 1)$, cf. [CF79]. The optimal value of that exponent is known in dimension $n = 1$ ($\alpha = \min\{1, 1/(m-1)\}$) but in higher dimensions is still under study.

Subsections 4.1.3, 4.1.4 The Harnack inequality in the fast diffusion range is part of a paper with M. Bonforte [BV05]. Earlier results were due to [DK92].

Harnack inequalities for degenerate parabolic problems are treated in many references, both for linear and nonlinear equations, cf. [ChS84, DiB93, Sk98, AP04], but the theory is still incomplete.

Section 4.2 The results of this section seem to be new, even though they are based on standard properties.

Exercises

Exercise 4.1

(i) Write the lower estimate of formula (4.9) in the adimensional form

$$u(x_0, t) \geq c\,\frac{M_R(x_0)}{R^n}(t/t_*)^{-\alpha}. \tag{4.36}$$

(ii) Show by scaling considerations that any optimal formula will have the form

$$u(x_0, t) \geq (M_R(x_0)R^{-n})\, H(t/t_*). \tag{4.37}$$

(iii) Check that the heat equation formula (4.14) has such a form.
Hint for (ii): Use formula (4.8) with the choice $T = t_*$ and $r = R$ to reduce to the case $R = 1$ and $M = 1$ for u^*.

(iv) Calculate the bound that is obtained in the typical case where the initial mass of a Barenblatt solution is located at the boundary of the ball $B_R(0)$ and we estimate u at $x = 0$ for $t > 0$. Compare the plot for $m > 1$ with the plots obtained for $m = 1$ and $m < 1$.

Part II

Study of the subcritical FDE

The foregoing analysis gives a rather complete picture of the smoothing effects and decay rates for the PME, the HE, and also for the FDE when $m > m_c$, the so-called supercritical FD range. We will address in this part of the text the same type of problems in the case of the FDE with critical and subcritical exponents, $m \le m_c$. The picture turns out to be quite different, very rich both in physical phenomena and in analytical results with an interesting connection with geometry. The main novelties of this range are lack of regularity (solutions with integrable data need not be continuous, not even bounded) and extinction (solutions just disappear after a finite time). These phenomena cannot be avoided by any natural re-definition of the concept of solution. A severe form of extinction is extinction in zero time, which leads to plain non-existence of solutions. We will devote our main effort to understand these issues.

In the study of the FDE with subcritical m, it is convenient in a first approximation to impose the further condition $m > 0$ (this leaves us with nothing to discuss in dimensions $n = 1, 2$, but it is significant in dimensions $n \ge 3$). As we have pointed out in Section 1.2, the restriction $m > 0$ is important for the basic existence and uniqueness theory, hence our prudence. However, crossing the exponent $m = 0$ is not so important for some of the basic questions we want to address, like a priori estimates and extinction, as we will see. We recall that, in view of the desired extension to $m \le 0$ the equation is always written in the modified form $u_t = \nabla \cdot (u^{m-1} \nabla u)$.

When organizing the present material, it is advisable to relate it to the (m, p) plane. There appear in it two main dividing lines. One is the second critical value of the exponent, $m = 0$, just mentioned, that marks the passage to what we will call the *superfast FDE*. The other one, more obscure at this moment but quite important in practice, is the *critical line* $p = n(1 - m)/2$. Diagrams will pay a big role in the future, so we will present them here. Figure II.1 exhibits the different parameter regions that will be encountered in dimensions $n \ge 3$. The smoothing effects, just studied, correspond to the lined region for $m > m_c$. This region will be easily extended into the whole supercritical area $p > n(1 - m)/2$. Figure II.2 exhibits the diagrams in dimensions $n = 2$ and $n = 1$.

As the theory proceeds more exponents will appear. Thus, a third exponent $m = (n - 2)/(n + 2)$ (in dimensions $n \ge 3$), which is the inverse of the famous Sobolev exponent of the elliptic theory, will appear in Chapter 7 where we will explain the role it plays in our theory and the relation with elliptic problems.

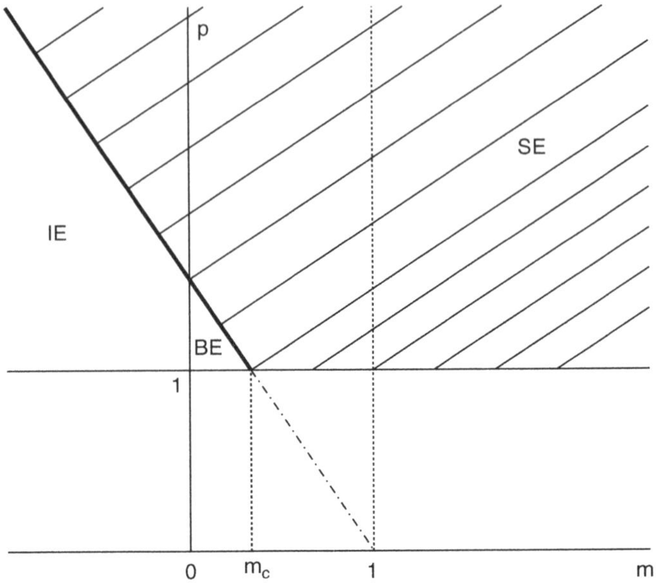

Figure II.1 The (m, p) diagram for the PME/FDE in dimensions $n \geq 3$. SE: smoothing effect, BE: backwards effect, IE: instantaneous extinction. Critical line $p = n(1-m)/2$ (in boldface).

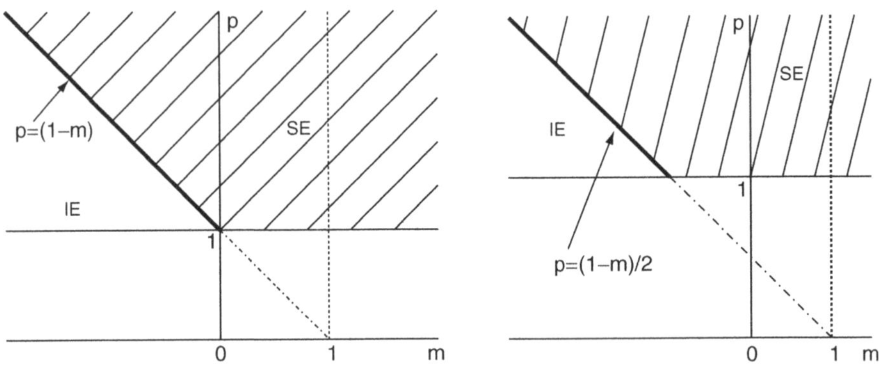

Figure II.2 Left: (m, p) diagram for the PME/FDE in dimension $n = 2$. Right: (m, p) diagram for the PME/FDE in dimension $n = 1$.

Here is an overview of this part: the phenomenon of extinction is discussed in Chapter 5. A simple and beautiful application of the Marcinkiewicz spaces is presented. The backward effect (L^p into L^1) is covered in Section 5.4; it happens only for $m > 0$ and $n \geq 3$; to our knowledge, it is described for the first time.

The analysis of the critical line is resumed in Chapter 6 and the phenomenon of delayed regularity is described; this is another major contribution of the text and it gives sharp information about the functional properties of the FDE semigroup.

Chapter 7 deals with the analysis of large-time behaviour where we give preference to L^1 data. In this case we cover important results taken from other sources plus our contributions. As m goes down to zero the phenomenon of extinction in finite time appears related to an increasingly slower diffusion strength of the process; for $m < m_s = (n-2)/(n+2)$ we can even talk of relative concentration.

Chapter 8 studies two transition situations where non-uniqueness plays an important role. The first deals with the range $-1 < m \le 0$ in $n = 1$, which looks like supercritical but contains the non-uniqueness phenomenon for the Cauchy problem. The second is the study of logarithmic diffusion in the plane, i.e., the case $m = 0$ for $n = 2$ which has many appealing features for the analyst and the geometer.

We then examine the special results for superfast diffusion, i.e., the range $m \le 0$ in Chapter 9. Instantaneous extinction is discussed in Section 9.2. The question of local estimates is introduced in Section 9.3.

Section 10 contains a commented summary of the main results obtained for the PME and FDE. We are adding the discussion on the existence of source solutions in the whole range $m \in \mathbf{R}$; as a key tool in that analysis we introduce the concept of source solution with background.

5

Subcritical range of the FDE. Critical line. Extinction. Backward effect

We begin here the study of the smoothing effect and decay rates for the FDE in the subcritical range $m < m_c$, and also for the critical exponent $m_c = (n-2)/n$. While advancing some of the results, which are valid for the whole subcritical range $m < m_c$, we will concentrate on the case $m > 0$; a detailed analysis of the lower subrange $m \le 0$ is delayed to Chapters 9 *et seq.*

Here is a summary of the chapter: after some preliminaries, in which the concept of critical line is introduced, Section 5.2 contains the proof of extinction of solutions in the Marcinkiewicz space $M^{p_*}(\mathbf{R}^n)$, that we characterize as the natural extinction space among all the spaces $M^p(\mathbf{R}^n)$ and $L^p(\mathbf{R}^n)$. An explicit separate-variable solution is now the model on which we apply the worst-case strategy. Interesting to notice that this solution is a close relative of the ZKB solutions that we used for $m > m_c$.

We also devote some space to the question of necessary conditions and the continuity of the extinction time T as a function of u_0. This is covered in Section 5.3. We point out at this moment that even if extinction in finite time is the main feature of this exponent range, not all solutions with data in an L^p space extinguish in finite time unless p takes a special value, $p = p_*$.

Then, we proceed in Section 5.4 with the construction of the global self-similar solutions that increase their space decay rate for all positive times. This is the technical basis for the proof of the backward smoothing effect (L^p into L^1) that happens below the critical line of the (m, p)-plane of Figure II.1.

We devote Section 5.5 to explaining how and where the whole mass is lost in the process of extinction. The conclusion is that mass is 'lost to infinity', a manifestation of the very fast diffusion process for small densities. We give an explanation in terms of particles escaping to infinity in finite time. This property is typical of the FD range $m < m_c$.

We return in Section 5.6 to the study of forward effects when dealing with the critical exponent $m = m_c$ with starting space L^1. There, we examine the absence of smoothing effect L^1–L^∞, the existence of weak effects is discussed, as well as exponential rates of asymptotic decay.

Finally, Section 5.7 discusses extinction as a form of blow-up after a suitable change of variables. Blow-up is a major topic in the study of evolution equations.

5.1 Preliminaries. Critical line

Significant novelties in the mathematical theory of the FDE appear once we cross the line $m = m_c$ downwards. Thus, concerning the basic problem of optimal space for existence, Brezis and Friedman proved in [BF83] that there can be no solution of the equation if $m \le m_c$ when the initial data is a Dirac mass, so that we lose our main example for the comparison theory, the source solution. Pierre [Pi87] extended the non-existence result to measures supported in sets of small capacity if $m < m_c$. But at least weak solutions exist for all initial data $u_0 \in L^1_{\rm loc}(\mathbf{R}^n)$ when $0 < m < m_c$, and moreover they are global in time, $u \in C([0,\infty) : L^1_{\rm loc}(\mathbf{R}^n))$. Such an existence result is not guaranteed when $m \le 0$, as we have already mentioned and we will see in detail in Section 9.

We will see in this and following chapters solutions with peculiar properties, like lack of regularization and extinction. It could be thought that this is the product of non-uniqueness and happens for some strange initial data or some strange class of solutions. This is not the case here. Herrero and Pierre proved in [HP85] a strong uniqueness theorem for general initial data in $L^1_{\rm loc}(\mathbf{R}^n)$. Other strong uniqueness results appear in [BrC79] and [Pi82]; they ensure that we are working all the time in classes of unique weak solutions, which are limits of bounded smooth solutions of the same equation. Moreover, the peculiar properties happen to solutions living in standard L^p spaces for different values of p.

A very important role is played by the property of infinite speed of propagation. The regularity theory says that in this range and at every fixed time, non-negative bounded solutions are either positive and smooth everywhere or identically zero. Once the solution touches the level zero at a time t_1, then it continues as identically zero for all $t \ge t_1$. One may wonder what is the meaning of such a solution, but there are satisfying answers: such an object is the limit of the smooth approximations constructed by regularizing the data (and the equation if needed).

However, and contrary to the 'upper' fast diffusion range $m_c < m < 1$, it is not true that locally integrable data produce locally bounded solutions at positive times, as the various examples to be constructed below will show. Attention must be paid therefore to the existence and properties of large classes of weak solutions that are not smooth, not even locally bounded. This leads to a main fact to be pointed out about smoothing estimates: there can be no L^1–L^∞ effects. Indeed, they cannot exist for a number of reasons. To begin with, after the result of Brezis and Friedman there is no worst-case solution to copy from in this setting.

A fundamental objection is the following: a universal estimate like (2.12) (a strong smoothing effect) would have exponents given by the same algebraic formulas, as a consequence of the scaling arguments already explained. But for $m < m_c$ we have a time exponent $-\alpha > 0$, and this means that the solution would increase in L^∞ norm in time like $t^{-\alpha}$, a contradiction with the properties of standard classes of solutions (by the maximum principle). Note finally that the best constant $c(m,n)$ of formula (2.14) goes to infinity as $n(m-1)+2 \to 0$.

We will see in a moment that the absence of smoothing into L^∞, as well as a new phenomenon, called extinction in finite time, happen for the FDE precisely when the starting space has exponent p for which the similarity exponents α and σ cannot be calculated by formula (3.2), i.e., for $n(m-1)+2p=0$. This is explicitly demonstrated with special solutions that play a key role in this chapter. This condition is therefore the consequence of an algebraic calculation based on the scaling of the equation. We are led to consider the following definition.

Definition 5.1 The line $mn+2p=n$ in the (m,p)-plane is called the *critical line*. The position of this important line in the plane depends strongly on the dimension (see Figure II.1 for $n\ge 3$). We are interested of course in values of $p\ge 1$. In a first approach we are also assuming that $m>0$. In that case there is no useful part of the critical line in dimensions $n=1,2$. The value of the function space exponent is

$$p_* = n(1-m)/2, \tag{5.1}$$

which is larger than 1 for $m<m_c$. It will be called the critical exponent, or 'the exponent on the critical line'.

5.1.1 Smoothing effects above the critical line

Let us first consolidate the ground that stays firm from the analysis done for $m>m_c$. Standard smoothing from M^p into L^∞ occurs for p above the critical line, $p>p_*$ (so that $p>1$ since $m\le m_c$, see the diagrams) and the proof is obtained just as in Chapter 3 by comparison with the self-similar solution $U(x,t)$ with data $u_0(x)=A\,|x|^{-n/p}$ and $p>p_*$. The effects are indeed true for $u_0\in M^p(\mathbf{R}^n)$ with $p>p_*(m,n)$ and $0<m\le m_c$. Again, the result is adapted to the case $m\le 0$ without difficulty. As a conclusion, we have the standard strong smoothing effect into L^∞.

Theorem 5.1 *The results of Theorem* 3.4 *hold true even for* $m\le m_c$ *when* $u_0\in M^p(\mathbf{R}^n)$ *and* $p>p_*$, *i.e.,* $2p>n(1-m)$. *In particular,* $u(t)$ *is bounded for all* $t>0$.

Let us mention that L^p–L^∞ effect under the restriction to the range $p>p_*$ was proved by Di Benedetto and Kwong [DK92]. The attempt to improve that result into the critical exponent $p=p_*$ will lead to the striking results of delayed regularity in Chapter 6.

5.2 Extinction and the critical line

Extinction in finite time is the phenomenon whereby the evolution of some non-trivial initial data u_0 produces a non-trivial solution $u(t)$ in a time interval $0<t<T$ and then $u(t)\to 0$ as $t\nearrow T$. Usually, the convergence is uniform in x, but we only need in principle convergence in the wider space in which the theory is constructed, like $L^1_{\rm loc}(\mathbf{R}^n)$.

5.2.1 Solution in Marcinkiewicz space. Universal estimate

To begin our theory, we remark that for $m < m_c$ there is an explicit example of solution which completely vanishes in finite time, and has the form

$$U(x, t; T) = c_m \left(\frac{T - t}{|x|^2}\right)^{1/(1-m)} \tag{5.2}$$

with $T > 0$ arbitrary, and

$$c_m^{1-m} = k^{-1} = \frac{2n}{(m-1)\alpha} = 2\left(n - \frac{2}{1-m}\right). \tag{5.3}$$

Note that c_m can be defined as a real number because $(m-1)\alpha > 0$; note also that values $m \le 0$ are accepted here. This solution is the closest relative for $m < m_c$ of the ZKB solutions of the range $m > m_c$ passing through the IPSS formula (2.11). We will see the rest of the relatives in Exercise 5.2.

Formula (5.2) produces a weak solution of the FDE that is positive and smooth for $0 < t < T$ and $x \ne 0$; note further that it belongs to the Marcinkiewicz space M^{p_*} with $p_* = n(1-m)/2 > 1$. It is clear that U does not belong to a better space than M^{p_*} during its lifetime $0 < t < T$. In particular, (5.2) is not a bounded function and $U(t) \notin L^p_{\text{loc}}(\mathbf{R}^n)$ for any $p \ge p_*$, $0 < t < T$. Therefore, we cannot have a general improvement effect from $M^{p_*}(\mathbf{R}^n)$ into $L^q(\mathbf{R}^n)$ for any $q \ge p_*$. For the same reason, no improvement effect can start from $L^p(\mathbf{R}^n)$ with $1 < p < p_*$.

The positive result that we can derive from comparison with this solution is the following.

Theorem 5.2 *Let $n \ge 3$ and $m < m_c$. For every $u_0 \in M^{p_*}(\mathbf{R}^n)$, there exists a time $T > 0$ such that $u(x,t)$ vanishes for $t \ge T$. It can be estimated as*

$$0 < T \le d_m \, \|u_0\|_{M^p}^{1-m} := T_1(u_0), \tag{5.4}$$

where $d_m = (c_m \kappa_m)^{1-m}$ is a function of m and n. Moreover, for all $0 < t < T$ we have $u(t) \in M^{p_*}(\mathbf{R}^n)$ and $u^*(t) \prec U(t; T)$ where T is chosen so that $U(x, 0; T)$ has the same M^p-norm as u_0, i.e., $M = c_m \kappa_m T^{1-m}$.

Proof Constant κ_m is defined in (1.15) and c_m in (5.3). The proof is done by comparison with U, exactly as in Theorem 2.1, if we further assume that $u_0 \in L^1(\mathbf{R}^n)$. By density and continuous dependence, we can extend the comparison results of Theorem 1.3 to solutions in the Marcinkiewicz class. The result is adapted to the case $m \le 0$ without major difficulty using suitable approximation techniques that the theory provides. See Chapter 9 below for a careful treatment of that exponent range. □

5.2.2 Consequences

The theorem has an immediate and quite relevant consequence.

Corollary 5.3 *Under the assumptions of the theorem, the M^{p_*}-norm of the solution profile $u(t)$ is strictly decreasing in time, and we have the estimate*

$$\frac{d}{dt}\|u(t)\|_{M^{p_*}}^{1-m} \le -k. \tag{5.5}$$

The best constant $k = k(m,n)$ is attained by the explicit solution (5.2).

Proof The first statement is almost immediate but anyway is a consequence of the second. The derivative estimate at $t = 0$ follows once we see that the explicit solution satisfies: $\|U(t)\|_{M^{p_*}}^{1-m} = k(T-t)$ with $k = 1/d$. In order to obtain the derivative estimate for $t > 0$ we displace the origin of time. □

Formula (5.2) is a self-similar solution of Type II in the terminology to be introduced in Section 7.1.1. Such solutions play a prominent role in the study of the rates and patterns of extinction of that chapter. In fact, comparison with the model example (5.2) implies the remarkable formula (5.5), which is valid for all solutions $u \ge 0$ with data $u_0 \in M^{p_*}(\mathbf{R}^n)$ and does not depend in its formulation on any specific information on the data. It is therefore a *universal estimate*, a quite strong tool in the study of these equations, and in particular of extinction.

Number $k(m,n)$ express the minimum rate of dissipation of this class of solutions; the main point is that it is not zero! It seems to appear by magic. An easy computation gives for $k = 1/d$ the exact value

$$k(m,n) = 2\pi n^{1-m}\left(n - \frac{2}{1-m}\right)^m \Gamma\left(\frac{n}{2}+1\right)^{-2/n}. \tag{5.6}$$

which for $m = 0$ takes the simple form $k(0,n) = 2\pi n\Gamma(\frac{n}{2}+1)^{-2/n}$ with $n > 2$. Note that k goes to zero as $m \to m_c$.

An immediate consequence of the universal bound is the following lower bound of the way the solution approaches extinction.

Corollary 5.4 *Let $u \ge 0$ be a solution of the FDE with $0 < m < m_c$, $n \ge 3$, and let $T > 0$ be its extinction time. Then, there exists a constant $k(m,n) > 0$ such that*

$$\|u(t)\|_{M^{p_*}}^{1-m} \ge k(m,n)\,(T-t). \tag{5.7}$$

This inequality is optimal for data in M^{p_*}.

Remarks

(1) This is the simplest control from below on the rate of extinction. Theorem 5.2 with its universal estimate and Corollary 5.4 are basic results in the extinction study. So, they motivate some thoughts. First of all is the fact that (5.7) can be seen as a weak *Harnack inequality* near extinction. Note that with respect to the Harnack inequalities of the heat equation and porous medium equation, it is reversed, in the sense that it gives an integral *lower bound* at previous times in terms of information of what happens later. It goes also contrary to the smoothing effects of previous chapters, where bounds are from above.

We will discuss the matter of *extinction rates* in Chapter 7. We will show solutions that decrease in the L^∞ norm with very fast rates, we mean exponents of $(T-t)$, but they keep the above rate in the norm of M^{p_*}.

(2) In this parameter range of the FDE, focus has changed into the extinction phenomenon and our tool is the explicit extinction solution (5.2). We have noted that it is formally a close relative of the ZKB solutions (2.2). We only have to put $C = 0$ in the free constant and change t into $t - T$. We can say that they are the same algebraic family. The geometrical aspect and functional properties are however quite different!

(3) The relation of the space $M^{p_*}(\mathbf{R}^n)$ with extinction is stressed by the following scaling transformation:

$$\widetilde{u}(x,t) = k^{2/(1-m)} u(kx,t) \tag{5.8}$$

that keeps time invariant and conserves the M^p norm of the profile $u(\cdot,t)$ if and only if $p = p_*$. This means in practical terms that we can perform these (x,u) scalings to any solution without affecting its extinction time.

5.3 Some basic facts on extinction

Before proceeding with the main line of smoothing effects and extinction, we devote some space to explain a number of important facts on extinction that serve as background.

5.3.1 Necessary conditions for extinction

We may ask ourselves how natural the functional space $M^{p_*}(\mathbf{R}^n)$ is in this game. We now show that it is almost the correct space where extinction of non-negative solutions takes place. The restriction $0 < m < m_c$ is imposed, so that the results do not apply for $m \le 0$.

Lemma 5.5 *For every solution u of the fast diffusion equation with $0 < m < m_c$ that vanishes in a finite time $T > 0$, the initial data satisfy a growth condition of the form*

$$\int_{\{|x-x_0|\le R\}} u_0(x)\,dx \le c\,T^{1/(1-m)}|B_R|^{1/p'_*}, \qquad p'_* = p_*/(p_*-1), \tag{5.9}$$

for every $x_0 \in \mathbf{R}^n$ and every $R > 0$. The constant $c > 0$ depends only on m and n.

Proof We use the well-known local mass estimate by Herrero and Pierre [HP85] that says that there exists a constant $c = c(n,m) > 0$ such that for every weak solution $u(x,t) \ge 0$ of the FDE in $Q = \mathbf{R}^n \times (0,T)$ and every $0 < t, s < T$, $x_0 \in \mathbf{R}^n$ and $R > 0$ we have

$$\int_{\{|x-x_0|\le R\}} u(x,s)\,dx \le c\int_{\{|x-x_0|\le 2R\}} u(x,t)\,dx + c\,|t-s|^{\mu}\,R^{\sigma}, \tag{5.10}$$

with $\mu = 1/(1-m)$, $\sigma = n - (2/(1-m))$. Note that $\sigma > 0$ if and only if $m < m_c$.

Assume now that u has extinction time T. Take $t \to T$ in the above formula, so that $\int u(x,T)\,dx = 0$ in any ball, and let $s \to 0$. We get the stated condition. □

Note that condition (5.9) is a weak form of the definition of Marcinkiewicz space $M^{p_*}(\mathbf{R}^n)$. Indeed, we can introduce new spaces of the family of the Morrey spaces. Our choice is the following.

Definition 5.2 We define the space $\widetilde{M}^p(\mathbf{R}^n)$ as the set of $f \in L^1_{\rm loc}(\mathbf{R}^n)$ such that

$$\int_{B_R(x_0)} |f(x)|\,dx \le C\,|R|^{n(p-1)/p} \quad \forall x_0 \in \mathbf{R}^n,\ R > 0,$$

with norm $|||f|||_p$ given by the minimum of the C in the above expression.

In the notation of Appendix AI.3, our $\widetilde{M}^p(\mathbf{R}^n)$ is called $\widetilde{M}^{1,\lambda}(\mathbf{R}^n)$ with $\lambda = n/p'$, $p' = p/(p-1)$. We have $M^p(\mathbf{R}^n) \subset \widetilde{M}^p(\mathbf{R}^n)$, and the norms are the same for rearranged functions. We then have:

Theorem 5.6 *If a solution u of the FDE vanishes in finite time T, then $u(t) \in \widetilde{M}^{p_*}(\mathbf{R}^n)$ for all $t < T$ and there is a $c = c(m,n)$ such that*

$$|||u(t)|||_{p_*} \le c(T-t)^{1/(1-m)}. \tag{5.11}$$

Hence, $u(t) \to 0$ in $L^p_{\rm loc}$ as $t \to T$ for all $p < p_*$. If moreover $u_0 \in L^1(\mathbf{R}^n)$, then $u(t) \to 0$ in $L^1(\mathbf{R}^n)$ as $t \to T$.

Proof We only need to prove the last result, the rest is easy. Since we have convergence in $L^1_{\rm loc}$, the result holds if we have a uniform bound on exterior domains

$$\int_{\{|x|>R\}} u(x,t)\,dx \le \varepsilon(R) \tag{5.12}$$

with $\varepsilon(R) \to 0$ as $R \to \infty$ uniformly in $0 < t < T$.

Now, when u_0 is bounded and compactly supported, we can bound it above by a function of the form $\tilde{u}_0(x) = A|x|^{-\gamma}$ with $\gamma \in (2/(1-m), n)$. According to the theory that we will develop in Section 5.4, the solution $\tilde{u}(x,t)$ lives for $0 < t < T$. By comparison, the same applies to u. When u_0 is a more general function in $L^1(\mathbb{R}^n)$, we use approximation of the initial data and L^1 contraction of the solution. We conclude that $u(t)$ is uniformly small in L^1-norm on exterior domains. □

The previous estimate implies in particular a bound from below of the extinction time in terms of the local L^1 norms of the initial data of the form

$$\int_{B_R(x_0)} u_0(x)\,dx \le C\,T^{1/(1-m)}\,R^{n(1-p_*)/p_*}, \tag{5.13}$$

which is valid for every solution u with extinction time T and every x_0 and R. This estimate complements the upper estimate (5.4).

5.3.2 Extinction spaces

In order to clarify the issue of extinction we propose the following definition.

Definition 5.3 A normed space $X \in L^1_{\rm loc}(\mathbf{R}^n)$ is called an *extinction space* for the FDE with exponent m if for every $u_0 \in X$ there exists a solution of the FDE flow that vanishes in a finite time $T(u_0)$. If moreover T can be bounded above in terms of the norm of u_0 in X, we say that X is an *adapted extinction space*.

We have the following result

Theorem 5.7 *$M^{p_*}(\mathbf{R}^n)$ is an adapted extinction space for the FDE if $m < m_c$, and every subspace is too. The spaces $L^p(\mathbf{R}^n)$ and $M^p(\mathbf{R}^n)$ with $p > p_*$ are not extinction spaces. Neither are $L^p(\mathbf{R}^n)$ and $M^p(\mathbf{R}^n)$ when $1 \le p < p_*$ if $m > 0$.*

The positive statement is already proved. The negative one will be the consequence of the construction of solutions in those L^p-spaces that exist in a non-trivial way for infinite time.

(i) The case $p > p_*$ is a consequence of the results of Subsection 5.1.1: the self-similar solution $U(x,t)$ mentioned there is bounded for all $t > 0$ and keeps the behaviour $U(x,t) \sim C(t)|x|^{-n/p}$ for all times. This means that we have a global-in-time solution that lives in the space $L^{p'}(\mathbf{R}^n)$ for all $p' > p$. By considering the solution $u(x,t) = U(x,t+\tau)$ with $\tau > 0$ we may assume that $u_0 \in L^{p'}(\mathbf{R}^n)$. It follows that $L^{p'}(\mathbf{R}^n)$ is not an extinction space for $p' > p_*$.

(ii) In the case $p < p_*$, $m > 0$, suitable solutions are constructed in Subsection 5.4.1, cf. Theorems 5.11, 5.14, as part of the study of the backward smoothing effect.

Corollary 5.8 *Let $0 < m < m_c$. For every $p \ge 1$ and $p \ne p_*$ there exist functions $u_0 \in L^p(\mathbf{R}^n)$, $u_0 > 0$, such that the corresponding solution of the FDE exists globally in time. Moreover, the solutions with $p < p_*$ are unbounded for every $t > 0$. For $p > p_*$ such solutions are bounded.*

We ask the reader to complete the proof using the proof of Theorem 5.7 and Exercise 5.14. A very simple example to show that the L^1-norm behaves badly with respect to estimating the extinction time is given in Exercise 5.1. We invite the reader to do that homework.

5.3.3 Continuous dependence of the extinction time

One may wonder if there is also a bound from above for the extinction time in terms of an L^1-norm. It is clear that this can only hold if the global norm is taken into account. Recall that we already have such an estimate in terms of the M^{p_*}-norm, formula (5.3). In fact, we pose a deeper question: to prove some dependence of the extinction time on the initial data in L^1-norm. Here is a partial result.

Theorem 5.9 *Let us consider the solutions of the FDE with $0 < m < m_c$ and $u_0 \in L^1(\mathbf{R}^n) \cap L^\infty(\mathbf{R}^n)$. Then, the extinction time $T(u)$ is a continuous function of u_0 in the*

$L^1(\mathbf{R}^n)$ norm if we act on bounded subsets of $L^\infty(\mathbf{R}^n)$. Actually, there is a constant $c = c(m,n)$ such that for two solutions u_1, u_2 we have

$$|T(u_2) - T(u_1)| \le c\|u_{0,1} - u_{0,2}\|_1^{2/n} N^{1-m-(2/n)}, \tag{5.14}$$

where $N = \max\{\|u_{0,1}\|_\infty, \|u_{0,2}\|_\infty\}$.

Proof (i) We may assume without lack of generality that $u_{01} \le u_{02}$, so that $T_1 \le T_2$. The general result will then hold by a comparison argument.

(ii) We know that the solutions are continuous functions of time with respect to the L^1-norm, and in fact the map: $u_0 \mapsto u(t)$ is a contraction in that norm. We want to control $T_2 - T_1$ in terms of $\|u_{0,2} - u_{0,1}\|_1 \le \varepsilon$ and also $\|u_{0,2}\|_\infty$, $\|u_{0,1}\|_\infty$. At $t = T_1$ we have $u_1(T_1) = 0$, hence

$$\|u_2(T_1)\|_1 \le \varepsilon, \qquad \|u_2(T_1)\|_\infty \le N_2 = \|u_{0,2}\|_\infty.$$

(iii) Let us prove that under these assumptions $\|u_2(T_1)\|_{M^{p_*}} \le \delta$ with δ small. We have the inequalities

$$\int_K u_2(t)\,dx \le \|u_2(t)\|_1, \qquad \int_K u_2(t)\,dx \le \|u_2(t)\|_\infty |K|.$$

The result holds if for each K one of these two options is true:

$$\varepsilon \le \delta|K|^q, \quad N_2|K| \le \delta|K|^q \tag{5.15}$$

with $q = 1 - (1/p_*)$. We select the first when $|K|$ is large and the second when it is small. The optimal dependence of δ on ε is obtained using as dividing value $|K| = c_1\varepsilon/N_2$, and then

$$\delta = c_2 \varepsilon^{1/p_*} N_2^{(p_*-1)/p_*}.$$

(iv) Once (5.15) is proved, the conclusion follows from Theorem 5.2. We get

$$T_2 - T_1 \le c\,\delta^{1-m}.$$

This proves the result. □

Remarks

(1) Putting $u_1 = 0$ we get the desired bound from above for the extinction time in terms of $\|u_0\|_1$ and $\|u_0\|_\infty$.

(2) The condition of boundedness can be replaced by $u_0 \in L^p(\mathbf{R}^n)$ or $u_0 \in M^p(\mathbf{R}^n)$ with $p > p_*$ in view of the smoothing effect, Theorem 5.1. We ask the reader to write the dependence result that replaces (5.14). See Exercise 6.1 for an improvement of this result to data $u_0 \in L^{p_*}(\mathbf{R}^n)$.

(3) The result is false if posed in $L^1(\mathbf{R}^n)$ without any additional restriction on the size of the data in a convenient space. This is due to the concentration effects of the solutions with Dirac delta as initial data that imply that solutions with L^1 data may have very large waiting times with $\|u_0\|_1 \le 1$. See Exercise 5.1.

5.3.4 Dependence of the extinction time on m

We can use the comparison of solutions of equations with different diffusivities (see Theorem 1.3) to get the following result.

Proposition 5.10 *Let $u_0 \in L^1(\mathbf{R}^n)$, $u_0 \le 1$, and let u_m be the solution of the FDE with exponent $m \in (0, m_c)$. Then, the extinction time $T_m(u_0)$ is a monotone increasing function of m.*

We are not excluding the possibility that $T_m(u_0) = \infty$ for some or all m. The result is not true if u is not bounded above by 1, but then the times depend monotonically up to a constant. See Exercise 5.5.

5.4 The fast-diffusion backward effect

We have no hope of a forward smoothing effect when p lies below the critical line in the (m, p) plane (i.e., when $p < p_*(m)$), at least in its strong form, because of the scaling argument already explained. However, the algebra of similarity exponents leads to an important novelty, the *backward embedding*, which is the content of the present section. Again, the analysis of the phenomenon is intimately tied to the properties of some special solutions which we study next.

5.4.1 Some self-similar solutions

We want to find bounds for the solutions of the FDE with initial data in the low L^p spaces, $1 < p < p_*$ when $0 < m < m_c = (n-2)/n$ and $n \ge 3$. The worst-case lies in the corresponding Marcinkiewicz space and is precisely given by the function

$$U_0(x) = A\,|x|^{-\gamma}, \tag{5.16}$$

with $\gamma = n/p$ and any $A > 0$. Therefore, γ lies in the range $2/(1-m) < \gamma < n$. Since function U_0 belongs to $L^1_{\rm loc}(\mathbf{R}^n)$, the established theory asserts the existence of a unique weak solution $u \in C([0,T) : L^1_{\rm loc}(\mathbf{R}^n))$ which is positive everywhere as long as it is defined (we are now well aware of the possibility of extinction in finite time!). Using the equation scaling as in Theorem 3.2 and Section 3.2.1, we conclude that the solution must have the self-similar form

$$U(x,t) = t^{-\alpha} f(|x|\,t^{-\beta}), \tag{5.17}$$

which in particular means that it will be defined for all $t > 0$. In order to comply with the initial data $U_0(x) = A\,|x|^{-\gamma}$, the two parameters must satisfy the condition $\alpha = \beta\gamma$. In order to satisfy the FDE equation, the self-similar exponents must be related by $2\beta = (1-m)\alpha + 1$. Both conditions imply the values

$$\beta = \frac{1}{2-\gamma(1-m)}, \qquad \alpha = \frac{\gamma}{2-\gamma(1-m)}. \tag{5.18}$$

Since in our case $\gamma > 2/(1-m) > 0$, we have $\beta, \alpha < 0$. Finally, the self-similar profile f as a function of the variable $\xi = |x|\, t^{-\beta}$ has to be a solution of the ODE:

$$\xi^{1-n}(\xi^{n-1} f^{m-1} f')' + \beta\xi f' + \alpha f = 0, \qquad \text{with } \xi > 0,\ f' := df/d\xi \tag{5.19}$$

(see Subsection 3.2.2). After these preparations, we may state the constructive result.

Theorem 5.11 *Let $m \in (0, m_c)$. For every $\gamma \in (2/(1-m), n)$ there exists a unique solution of the PME of the self-similar form* (5.17) *with initial data* (5.16). *It has exponents given by* (5.18), *and the profile function $f(\xi)$ solves equation* (5.19), *is singular as $\xi \to 0$ in the form $f(\xi) = A\xi^{-\gamma}(1 + O(1))$, and decreases to zero as $\xi \to \infty$.*

The proof is divided into several stages. The behaviour as $\xi \to \infty$, which is quite relevant for our purposes, will be given separately in Theorem 5.14 below.

Phase-plane analysis We know that the solution exists and has this self-similar form. We need to study the behaviour of the profile f. General properties indicate that it is positive, monotone decreasing and tends to zero at infinity. In order to know more, we perform the analysis after introducing the following variables:

$$\xi = e^r, \qquad X(r) = \frac{\xi f'}{f}, \qquad \text{and} \qquad Y(r) = \xi^2 f^{1-m}. \tag{5.20}$$

Then the functions X and Y satisfy the following autonomous system:

$$\begin{cases} \dot{X} = (2-n)X - mX^2 + (a + bX)Y, \\ \dot{Y} = (2 + (1-m)X)Y, \end{cases} \tag{5.21}$$

where $\dot{X} = dX/dr$, and we have written $a = -\alpha$, $b = -\beta$, so that $a, b > 0$. The reader should consult first Section 3.2 about the basic properties of this system. Let us notice to begin that there exist two critical points

$$O = (0, 0), \qquad A = ((2-n)/m, 0),$$

in the half-plane $Y \geq 0$.

Let us identify the profile of the solution we are studying. It is positive, so we only need to consider orbits where $\{Y > 0\}$. Since it is monotone decreasing, we have $X < 0$. All together, we work in the quadrant $Q = \{X < 0,\ Y > 0\}$. As for end values, as $\xi \to 0$, i.e., $r \to -\infty$ we will have $f(\xi) \sim A\,\xi^{-\gamma}$, hence

$$X(-\infty) = -\gamma, \qquad Y(r) \sim \xi^{-\gamma(1-m)+2} \to +\infty$$

as $r \to -\infty$. We explain in Subsection 3.2.3 how a change of parametrization of the orbit $(X(r), Y(r))$ allows us to obtain profiles with all different constants $A > 0$ in the behaviour as $r \to -\infty$ from only one orbit.

We are interested in knowing what happens as $\xi, r \to \infty$.

Critical points The second line of (5.21) selects the values $Y = 0$ and $X = -2/(1-m)$. For $Y = 0$ there exist two critical points: $O = (0,0)$ and $A = ((2-n)/m, 0)$. They lie on the border of the correct quadrant. The second option for obtaining a critical point was $X = -2/(1-m)$, and then the first line gives

$$Y = \frac{2\,(2 - n(1-m))}{1-m} < 0,$$

which falls out of the quadrant. We also need the linearized analysis around the two critical points that lie on the border of Q.

Proposition 5.12 *The linearization of* (5.21) *around* $O = (0,0)$ *has matrix*

$$\begin{pmatrix} 2-n & a \\ 0 & 2 \end{pmatrix}$$

with eigenvalues $\lambda_1 = -(n-2)$ *and* $\lambda_2 = 2$ *and corresponding eigenvectors* $\vec{e_1} = (1,0)$ *and* $\vec{e_2} = (a, n)$. *Thus, O is a saddle point.*

Growth in Q After reading Section 3.2 if necessary, we see that Y grows as long as $X > c = -2/(1-m)$, a vertical line which lies to the right of A since $n(1-m) > 2$. It decreases for $X < c$. The growth of X is not so simple, but we know that $\dot{X} > 0$ near the axis $X = 0$. Indeed, at $X = 0$ we have

$$\dot{X} = aY, \quad \dot{Y} = 2Y,$$

which means that we have outgoing flux on the border $X = 0$ of Q. These orbits enter the region $X > 0$ of increasing profiles f, that do not match our orbit. Note also that on the line $Y = 0$ the flow is tangent and goes towards the origin in the segment $\overline{AO}$, towards minus infinity when $X < (2-n)/m$.

Indeed, X grows in Q to the right of the 'line of vertical directions',

$$Y = f_v(X) = \frac{X(mX + n - 2)}{a + bX},$$

which passes through the point $A = (-(n-2)/m, 0)$ and has a vertical asymptote at $X = -a/b = -\gamma$. Note that $-(n-2)/m < -\gamma < -2/(1-m)$. The orbit we are looking for comes precisely down from that vertical asymptote.

Choosing the correct orbit The phase-plane analysis leads so far to the conclusion that the orbit we are looking for must either end in A or approach the axis $Y = 0$ with $X \to -\infty$. Now, in the last case, it will have blow up for X at a finite distance, since asymptotically $\dot{X} \sim -m\,X^2$ when $X \to -\infty$. This means that $f(\xi_1) = 0$ for some finite $\xi_1 > 0$. It cannot be the case in our orbit.

We conclude that the connection that we are looking for must end in A, hence as $r \to \infty$ we have $X \to -(n-2)/m$, which means that

$$f(\xi) \sim \xi^{-(n-2)/m} \quad \text{as } \xi \to \infty.$$

Since $n(1-m) > 2$ this is integrable at infinity. The construction is complete.

As a curiosity, we give the linearized analysis around the critical point A where the orbit ends.

Proposition 5.13 *Let $m > 0$, $n > 2$. The linearization of (5.21) around $A = ((2-n)/m, 0)$ has matrix*

$$\begin{pmatrix} n-2 & -\frac{\alpha m+(2-n)\beta}{m} \\ 0 & \frac{n(m-1)+2}{m} \end{pmatrix}$$

with eigenvalues $\lambda_1 = n - 2 > 0$ and $\lambda_2 = (n(m-1)+2)/m < 0$ and corresponding eigenvectors $e_1 = (1, 0)$ and $e_2 = (c, 1)$ with $c > 0$. A is a saddle point.

The data of Figure 5.1 are: $n = 3$, $m = 1/6$, $\gamma = 5/2$. They satisfy $n(1-m) > 2$ and $2/(1-m) < \gamma < n$. Then, $p = n/\gamma = 6/5 < p_* = n(1-m)/2 = 5/4$. Besides,

$$a = \frac{\gamma}{\gamma(1-m)-2} = 30, \quad b = \frac{1}{\gamma(1-m)-2} = 12.$$

The computation focuses on the region $X \le 0$, $Y \ge 0$. The orbit we look for, as the desired self-similar solution, arrives at the saddle point $(n-2)/m, 0) = (-6, 0)$ coming from the asymptote $X = -\gamma$. There are more curves to the right, they start at the asymptote and cross the vertical axis to get out of the quadrant. To the left, they go to $Y = 0$, $X \to -\infty$. Summing up, we have

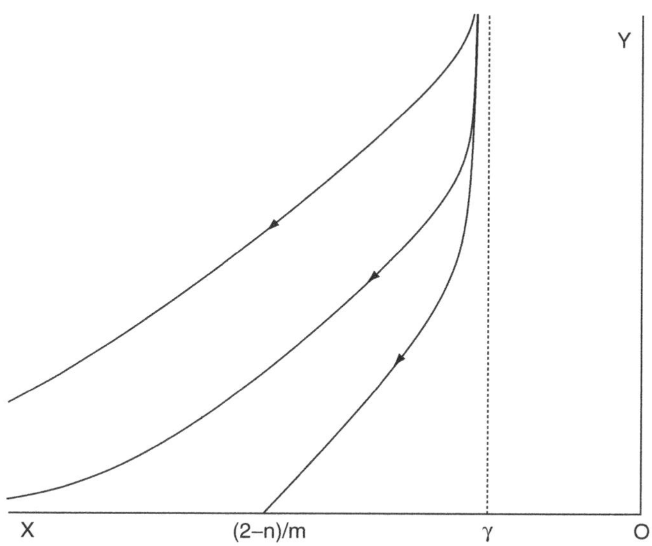

Figure 5.1 Phase-plane graph of the self-similar solutions for the backwards effect.

Theorem 5.14 *The unique self-similar profile corresponding to the solution with initial data* (5.16) *behaves at infinity like*

$$f(\xi) \sim c\xi^{-(n-2)/m}. \tag{5.22}$$

Therefore, $U(x,t)$ *is integrable in* $\mathbf{R}^n$ *for all* $t > 0$

$$\int_{\mathbf{R}^n} U(x,t)\,dx < \infty. \tag{5.23}$$

Moreover, The solution has for fixed $t > 0$ *the asymptotic behaviour*

$$U^m(x,t) \sim \frac{t^\sigma}{|x|^{n-2}}, \qquad |x| \to \infty,$$

with

$$\sigma = \beta(n-2) - \alpha m = b(m\gamma - (n-2)) = \frac{n(m-p)+2p}{n(1-m)-2p} < 0.$$

Note also that the special orbit satisfies the bounds $-(n-2)/m < X(r) < -\gamma$. Using formula (5.17) we conclude that

$$u_t = bt^{a-1} f(\xi)(X+\gamma) < 0,$$

which shows that $u_t < 0$ everywhere.

5.4.2 The backward estimates

We can now formulate the *backward embedding* result. By this we mean that data $u_0 \in M^p(\mathbf{R}^n)$ (in particular, $L^p(\mathbf{R}^n)$) produce solutions $u(t) \in L^1(\mathbf{R}^n)$ for all $t > 0$.

Theorem 5.15 *Let* $n \geq 3$, $0 < m < m_c$ *and* $1 < p < p_*$. *For every* $u_0 \in M^p(\mathbf{R}^n)$, *and every* $t > 0$ *we have* $u(t) \in L^1(\mathbf{R}^n)$ *and*

$$\|u(x,t)\|_1 \leq c(m,n,p,1)\,\|u_0\|_{M^p}^{\sigma(p)}\, t^{-\alpha(p)} \tag{5.24}$$

with

$$\alpha(p) = \frac{n(p-1)}{n(1-m)-2p}, \qquad \sigma(p) = \frac{p\,(n(1-m)-2)}{n(1-m)-2p}. \tag{5.25}$$

With these formulas, L^q-decay as $t \to \infty$ occurs whenever $1 \leq q < p < p_*$. We leave it as an exercise for the reader to prove the general formula by interpolation of the previous cases.

Remark Being used by now to the forward effect, it may look at first sight that the backward effect (BE) implies no improvement in the knowledge of the function, and this is true at the local level. The significance of the BE is understood when we realize that the information we get corresponds to the better behaviour as $|x| \to \infty$. We will add some illuminating information on the significance of this improved decay and its relation with the standard smoothing in Section 7.5 when treating the Yamabe flow.

Basin of attraction into $L^1(\mathbf{R}^n)$ The backwards smoothing effect can be stated for functional analysis buffs in the following way:

Functional version of the BE *Data in the spaces $M^p(\mathbf{R}^n)$ are 'attracted' into $L^1(\mathbf{R}^n)$ for positive times of the FDE evolution if $0 < m < m_c$ and $1 < p < p_*$.*

See in this respect the open problem posed at the end of the chapter.

5.4.3 An excursion into extended theories

Once started, the preceding phase-plane analysis has a certain life of its own and leads to some novel directions and results. Let us briefly present one of them.

The type of connection in phase plane we have been looking for exists whenever $\gamma > 2/(1-m)$ and the existence has nothing to do with the bound $\gamma < n$. This limitation is equivalent to choosing locally integrable self-similar solutions $u(t) \in L^1_{\text{loc}}(\mathbf{R}^n)$ for all t, a reasonable assumption for the existence theory, but not strictly needed.

We can easily see that when $\gamma \geq n$ the same type of connection exists and still enters as $\xi \to \infty$ the point A. It corresponds to a self-similar solution defined for $x \neq 0$ and all $t \geq 0$ with initial data of form

$$U(x,0) = A\,|x|^{-\gamma}, \tag{5.26}$$

For $t > 0$ the solution $U(x,t)$ has this precise behaviour as $|x| \to 0$, which is non-integrable for $\gamma \geq n$; on the other hand, it always has an integrable behaviour as $|x| \to \infty$. Therefore, we have a strong form of the BE.

The theory of solutions with non-integrable initial singularities that stay singular in time has been completely developed for the FDE in the upper range $m_c < m < 1$ by Chasseigne and the author in [ChV02] under the name of *extended continuous solutions with strong singularities*. However, the theory of such solutions for $m \leq m_c$ is still partially understood. Since this is a diversion from the main subject, we will refrain from pushing more the topic and propose some exercises to the interested reader.

5.5 Explaining how mass is lost

Extinction is an important phenomenon in the theory of nonlinear diffusion that begs for an explanation. Next, we analyse the issue in different ways, using physical concepts of diffusion or particle approach.

5.5.1 Flux at infinity

We will try to provide a first explanation for solutions with integrable data. We will adopt the idea of considering the flow $u(x,t)$ as the evolution in time of a spatial mass distribution with density function $u(\cdot,t)$ and mass

$$M(t) = \int u(x,t)\,dx. \tag{5.27}$$

We have counted on the conservation of this aggregated quantity for $m \geq 1$, even in fast diffusion when $m \geq m_c$. A simple way of justifying that claim is to integrate in a ball of radius R at time t and obtain

$$\frac{d}{dt}\int_{|x|\leq R} u(x,t)\,dx = \int_{|x|=R} \frac{\partial u^m}{\partial n}\,dS = R^{n-1}\int_{|x|=1} \frac{\partial u^m}{\partial n}(Rx,t)\,d\sigma := \Phi(R,t),$$

where $\partial/\partial n$ indicates normal derivative and $d\sigma$ is the area element on $\mathbb{S}^{n-1}$. If u is radial, as the model solutions are, we have

$$\frac{d}{dt}\int_{|x|\leq R} u(x,t)\,dx = n\omega_n R^{n-1}\frac{\partial u^m(R,t)}{\partial r}.$$

Replacing the last derivative by the value in the model cases we see that it goes to zero as $R \to \infty$ whenever $m > m_c$. Indeed, for $m_c < m < 1$ we have $u(r,t) \sim r^{-2/(1-m)}$. But the calculation with the same type of asymptotic behaviour gives an infinite flux as $R \to \infty$ in case $m < m_c$. It means that either the asymptotic behaviour as $|x| \to \infty$ is different, or the total mass is not conserved, or both. In fact, the latter is true: some mass is lost as time grows.

The argument of conservation of mass can be extended to general solutions in case $m > m_c$ by using for instance the Herrero–Pierre inequality (3.26)

$$\int_{\{|x-x_0|\leq R\}} u(x,s)\,dx \leq c\int_{\{|x-x_0|\leq 2R\}} u(x,t)\,dx + c\,|t-s|^{\mu}\,R^{\sigma},$$

with $\mu = 1/(1-m)$, $\sigma = n - (2/(1-m))$. Since exponent σ is negative we may let $R \to \infty$ for fixed s and t and conclude that the total mass is constant (if it is finite). On the other hand, $\sigma > 0$ for $m < m_c$ and loss of mass is possible. Indeed, in this case we have seen that the solutions with minimal behaviour have a first-order expansion $u(r,t) \sim c(t)r^{(2-n)/m}$. This formally gives an outgoing flux at $r = R$

$$\Phi(R,t) \sim -c(t)n\omega_n, \tag{5.28}$$

which means in the limit that $M'(t) = -c(t)n\omega_n$, as verified in practice for the self-similar solutions to be constructed in Chapter 9. On the other hand, the solutions with behaviour $u \sim c(t)r^{-2/(1-m)}$ as $r \to \infty$ have both infinite mass and infinite outgoing flux. So the main question is, where is all this lost mass going?

5.5.2 *Mass goes to infinity*

An answer is provided by the approximation with strictly positive data,

$$u_{0\varepsilon}(x) = u_0(x) + \varepsilon. \tag{5.29}$$

We first observe that the solution with these data, $u_\varepsilon(x,t)$, exists for all time and that $v_\varepsilon(x,t) = u_\varepsilon(x,t) - \varepsilon$ is an integrable distribution that conserves mass (prove it).

Secondly, we can prove that $u_\varepsilon(x,t)$ converges in a monotone way to $u(x,t)$, and the convergence takes place in $L^1(Q_T)$, with $Q_T = B_R(0) \times (0,T)$, for every R and T. By interior regularity, the convergence is uniform in Q_T if u_0 is continuous. This means that

$$\lim_{\varepsilon\to 0} \int_{|x|\ge R} (u_\varepsilon(x,t) - \varepsilon)\, dx = M(0) - M(t).$$

This is where the lost mass is when you approximate: it lies in exterior sets of the form $|x| \ge R$ (i.e., neighbourhoods of infinity). The fact that mass is *lost to infinity* is a manifestation of the very fast diffusion process for small densities.

This approximation also shows that extinction solutions are limits of perfectly classical and bounded solutions which remain positive for all time.

5.5.3 Escape to infinity of particles

A very clear explanation of the phenomenon of mass loss by escape to infinity in the subcritical FDE comes from considering the solution $u \ge 0$ as the density of a substance composed of particles and the FDE as an instance of the general mass conservation law of continuum mechanics,

$$u_t + \nabla(u\,\mathbf{v}) = 0, \tag{5.30}$$

also called the transport equation. As explained in Appendix II, pages 207–210, the pointwise velocity of the particles is then defined as

$$\mathbf{v} = -u^{m-2}\nabla u = -\frac{1}{m-1}\nabla u^{m-1}. \tag{5.31}$$

The definition makes sense mathematically even if $m < 1$.

When we apply the particle approach to solutions having a specific asymptotic behaviour as $|x| \to \infty$, we have to integrate the kinematic equation for the trajectories $dx/dt = \mathbf{v}(x,t)$, where the vector field is known through the solution $u(x,t)$, to obtain the way in which particles that are initially located far away diverge to infinity with time. This needs precise information on the behaviour of $\mathbf{v}$.

(i) We explain in Appendix II.2 the behaviour for compactly supported solutions of the PME, the HE, and also the FDE in the range $m_c < m < 1$. The conclusion is that particles go to infinity with a power rate $x(t) \sim c\,t^\beta$. Actually, we explain that the result is tied to the curious linear dependence between velocity and position for the particles: $\mathbf{v} = \beta x/t$, which characterizes the ZKB solutions.

(ii) Let us now consider the same situation for the fast diffusion equation in the range $0 < m < m_c$. In this case, we have some solutions with the behaviour

$$u \sim c_m((T-t)/x^2)^{1/(1-m)},$$

$c_m^{m-1} = |\beta|(1-m)/2$, where now $\beta = 1/(2-(1-m)n) < 0$. We get a divergent behaviour for the speed

$$\mathbf{v} \sim |\beta| x/(T-t), \tag{5.32}$$

which looks like the ZKB behaviour but has $T-t$ instead of t. Integrating, we get

$$x(t) \sim x_0\,(T-t)^{-\beta}, \tag{5.33}$$

and the particles reach infinity at the extinction time T. Remember that in those cases the total mass is infinite.

(iii) Even more extreme is the situation when we consider for $0 < m < m_c$ the self-similar solutions describing the backward effect, cf. Theorem 5.14, which have minimal behaviour of the form

$$u^m(x,t) \sim \frac{t^\sigma}{|x|^{n-2}}, \qquad |x| \to \infty, \tag{5.34}$$

with $\sigma < 0$, or any other solutions with similar behaviour as $|x| \to \infty$. In that case the trajectories go to infinity at times between 0 and T in a way that can be explicitly integrated but gives a behaviour of the form

$$x(t) \sim c(t_* - t)^{-\mu}$$

for some $c, \mu > 0$ and $0 < t < t_* < T$. This explains precisely how mass is lost in a continuous way in the time interval $(0, T)$.

5.5.4 *Decay without apparent diffusion. An illuminating example of mass loss*

A convenient way to understand the phenomenon of 'loss of mass towards infinity' lies in the study of the source-type solutions. We take the limit case, $m = 0$, $n = 2$ that will be studied in Chapter 8, for which the mass loss effect is also true, and we approximate it with source-type solutions with $n = 2$ and $0 < m < 1$. They read:

$$U_m(x,t;M) = \left(\left(\frac{4\pi t}{M}\right)^{(1-m)/m} + \frac{(1-m)x^2}{4mt}\right)^{-\frac{1}{1-m}}. \tag{5.35}$$

We see that the limit $m \to 0$ is singular. In order to understand how that happens we perform some computations:

(i) Immediate inspection shows that $U_m(x,t) \le (M/4\pi t)^m \to 0$ as $m \to 0$ uniformly in the set $t \ge M/4\pi$. This is extinction in quite explicit form.

(ii) We compute the spread of the mass of the solution before extinction by integrating the function outside a given ball and we find the expression

$$M_m(R,t) := \int_{|x|\ge R} U_m(x,t)\,dx = 4\pi t \left[\left(\frac{4\pi t}{M}\right)^{\frac{1-m}{m}} + \frac{(1-m)R^2}{4mt}\right]^{-\frac{m}{1-m}}.$$

We have $0 \le M_m(R,t) \le M$. There is now a crucial observation: if we take the limit of this expression as $m \to 0$, it amounts precisely to $4\pi t$ as long as $4\pi t \le M$, and to M when $4\pi t \geq M$. The rate of mass loss per unit time, 4π, is a kind of magical number, independent of M. Summing up, we have

$$\lim_{m\to 0} \int_{B_R(0)} U_m(x,t;M)\,dx = (M - 4\pi t)_+,$$

whenever $0 < R < \infty$ is fixed. This shows the very precise rate in which the mass is lost. We will return to this computation in Subsection 8.2.1 when explaining the mass loss phenomenon in logarithmic diffusion. We propose to complete this computation in Exercise 5.14.

(iii) According to the trajectory analysis of Appendix II for the ZKB solutions, the trajectories of the particles are given by

$$x(t) = x_1\, t^{1/2m}.$$

In the limit the only possibility is that they stay at the origin for a time, in number enough for the mass to be $M - 4\pi t$, and then they jump instantaneously to $x = \infty$.

5.6 The end-point $m = m_c$

The FDE with critical exponent $m_c = (n-2)/n$ is somewhat special, as befits a borderline situation. The critical p is then $p_* = 1$. By Theorem 5.1 we know that the strong smoothing effect holds from all L^p, $p > 1$ into $L^\infty(\mathbf{R}^n)$. Let us examine here the case $p = 1$ in dimensions $n \geq 3$, so that $m_c > 0$. This case has not been discussed yet.

There is a main difference with the rest of the critical line. Namely, there is no extinction in this case; on the contrary, it is known that mass conservation holds, $\int u(x,t)\,dx = \int u_0(x)\,dx$, cf. [BC81].[1] Therefore, to every non-trivial initial data $u_0 \geq 0$ there corresponds a unique solution $u \in C([0,\infty) : L^1(\mathbf{R}^n))$ that conserves its mass globally in time. It is moreover proved that when the solution is locally bounded then $u \in C^\infty$ in (x,t).

Next, a novelty which may come as a surprise: since the strong smoothing effect is true for $m > m_c$, by taking some limit of the smoothing effects for $m > m_c$ we could expect that a smoothing effect from L^1 into L^∞ would be true with a decay rate faster than all negative powers of time maybe (since the exponent α in the smoothing effect tends to ∞ as $m \searrow m_c$. Actually, such an effect is not true in the present case even in its weak version. This contradicts the panorama of Chapter 6 where we will show the boundedness result on the critical line in the weak sense.

Let us recall that the situation of the critical end-point is different for dimensions $n = 1$ $(m_c = -1)$ and $n = 2$ $(m_c = 0)$ since we are then in the superfast diffusion range. See Section 9.4 and Chapter 8 below, respectively.

[1] The proof is as follows: mass conservation is true for $m > m_c$ and the solutions depend continuously on the nonlinearity in the norm of $C([0,\infty) : L^1(\mathbf{R}^n))$.

5.6.1 Exponential decay at the critical end-point

Exponential decay holds for the FDE with critical exponent m_c for a large number of initial data as we show next, but not for all L^1 data. This partial *weak effect with exponential rate* is the following

Theorem 5.16 *Let $m = m_c$ and $n \geq 3$. Solutions in $L^1(\mathbf{R}^n) \cap L^\infty(\mathbf{R}^n)$ decay in time according to the rate*

$$\log(1/\|u(t)\|_\infty) \sim C(n)M^{-2/(n-2)}t^{n/(n-2)} \tag{5.36}$$

as $t \to \infty$, $M = \|u_0\|_1$. This rate is sharp.

This decay was calculated by Galaktionov, Peletier and Vázquez in [GPV00], based on previous formal analysis of King [Ki93]. The proof uses a delicate technique of matched asymptotics that falls outside of the scope of the present work. We refer also to the book [GV03, Chapter 5] for this and related issues.

So far we have concluded that solutions with data in $L^1(\mathbf{R}^n) \cap L^\infty(\mathbf{R}^n)$ decay in time with a superexponential rate. On the other hand, the self-similar solutions with data in $M^P(\mathbf{R}^n)$ with $p > 1$ that serve us as comparison functions in Theorem 5.1 decay like a power $u(x,t) = O(t^{-\alpha_p})$ where α ranges from 0 to infinity. There are intermediate solutions with explicit exponential decay.

Example 5.1 Here is an explicit example:

$$u(x,t) = \frac{1}{(ax^2 + ke^{2nat})^{n/2}}, \tag{5.37}$$

with free parameters $a, k > 0$. This is one of the rare examples of explicit solution in this book that is not self-similar in the usual way. Note the we can get plain exponential decay

$$\|u(t)\|_\infty = O(e^{-n^2at})$$

with all possible coefficients $a > 0$. Notice that u is an *eternal solution* (i.e., it lives for $-\infty < t < \infty$), and that $u_t < 0$. Note also that for all $t \in \mathbf{R}$ we have $u(t) \notin L^1(\mathbf{R}^n)$, but $u(t) \in L^p(\mathbf{R}^n)$ for all $p > 1$. See more about this solution and its kinship with the ZKB family in Exercise 5.8.

5.6.2 No strong or weak smoothing

Let us now address the negative result. It has been proved in [ChV02] that even a weak smoothing effect from $L^1(\mathbf{R}^n)$ into $L^\infty(\mathbf{R}^n)$ does not hold. We prove here a stronger version of that result.

Theorem 5.17 *Let $m = m_c$ and $n \geq 3$. There is no smoothing effect, even in weak form, from $L^1(\mathbf{R}^n)$ into any $L^p(\mathbf{R}^n)$ with $p > 1$.*

The proof is based on the construction of some solution that does not have the required L^p-regularity for any p.

Example 5.2 There exists a positive function $u_0(x) \in L^1(\mathbf{R}^n)$ such that the solution of equation $u_t = \Delta(u^m/m)$ with $m = (n-2)/n$, $n \geq 3$, is not in $L^p(\mathbf{R}^n)$ for any $p > 1$ and any $t > 0$. Moreover, $u(0,t) = \infty$ for all $t > 0$. We can construct one such solution that is smooth for $x \neq 0$, $t > 0$.

Proof (i) We start with a smooth and compactly supported initial function, $\varphi_1(x)$, that gives rise to a smooth and positive solution $u_1(x,t)$ of the equation which conserves the initial mass, say $\|u_1(t)\|_1 = \|\varphi_1\|_1 = 1$ for all time. It decays as $t \to \infty$ in the exponential manner described by (5.36).

Now, we obtain a collection of smooth solutions u_k with mass $1/k^2$ by means of the scaling transformation

$$v_k(x,t) = \frac{\lambda_k}{k^2} u_1(\lambda_k^{1/n}\, x, k^{4/n} t), \tag{5.38}$$

where λ_k will be chosen below as a steeply increasing function of k. In this way, for every $p \geq 1$ we have

$$\|v_k(t)\|_p = \lambda_k^{(p-1)/p} k^{-2} \|u_1(k^{4/n} t)\|_p.$$

We now fix $t = 1$ and take the sequence of norms $\|v_k(1)\|_p$, whose behaviour is known to us through the behaviour of u_1 for large times. The precise behaviour of u_1 is given by the estimate (see [GV03], page 171):

$$\log(1/u_1) \sim C_1(n) t^{n/(n-2)}$$

in a region of the form $|x| \leq R(t)$, with $\log R(t) \sim C_2(n)\, t^{n/(n-2)}$. Therefore,

$$\log \|u_1(t)\|_p \geq \left(\frac{C_2 n}{p} - C_1\right) t^{n/(n-2)}.$$

We get in any case a uniform estimate,

$$\|v_k(1)\|_p = \lambda_k^{(p-1)/p} k^{-2} \|u_1(k^{4/n})\|_p \geq C \lambda_k^{(p-1)/p} k^{-2} e^{-Ck^{4/(n-2)}}$$

with uniform constants $C = C(n)$. We only need to choose a highly increasing λ_k to make this expression tend to infinity as $k \to \infty$, no matter what is the value of $p > 1$.

(ii) Next, for the actual example we consider the solution $u(x,t)$ with data

$$u_0(x) = \sum_{k=1}^{\infty} v_k(x,0).$$

Clearly, $u_0 \in L^1(\mathbf{R}^n)$. Also, by comparison $u(x,t) \geq v_k(x,t)$ for every i. Therefore,

$$\|u(1)\|_p = \infty \quad \forall p > 1.$$

We translate this result into the interval $t \in (0,1)$ using the universal inequality $u_t \leq u/(1-m)t$ [BC81b], since it implies that $t \mapsto u(x,t)t^{-1/(1-m)}$ is a non-increasing function for a.e. fixed $x \in \mathbf{R}^n$.

(iii) We have proved the result for $0 \le t \le 1$. A scaling argument allows us to prove it for $0 \le t \le T$ for any T. By using a solution with data the sum of solutions of this form, an example valid for all times $0 \le t < \infty$ is obtained. We leave to the reader to complete those easy details.

(iv) If moreover, we choose ϕ_1 to be radially symmetric and non-increasing in $|x|$, then all the $v_k(t)$ will have the same property, hence so will u in the limit. The fact the $u(t)$ is integrable implies in that case that $u(x,t)$ is bounded on sets of the form $\{(x,t) : |x| \ge r, t_1 \le t \le t_2\}$. On the other hand, the divergence of the L^p norms implies that

$$\lim_{x\to 0} u(x,t) = \infty \quad \forall t > 0.$$

Finally, general theory implies that locally bounded non-negative solutions are indeed C^∞-smooth and positive in regions of boundedness. □

The idea of the construction is a variation of the one contained in paper [ChV02, Appendix A3], which considered the non-existence of the effect $L^1(\mathbf{R}^n)$–$L^\infty(\mathbf{R}^n)$. We cover here all spaces L^p with $p > 1$. This forces the present construction to be technically more involved than the one in [ChV02].

Example 5.3 By adding initial data of the previous form, given any countable set $S \subset \mathbf{R}^n$ we can construct a solution $u \in C([0,\infty) : L^1(\mathbf{R}^n))$ of the FDE with critical exponent $m = m_c$ that has a standing singularity at all points of the set S (standing singularity means that the solution is singular there for all times).

5.7 Extinction and blow-up

We devote this section to transform the extinction problem into a blow-up problem, i.e., one in which solutions tend to infinity in finite time. In view of the widespread interest in blow-up problems, and the advanced state of the studies in that field, the transformation can provide useful insights or produce interesting consequences of the extinction results.

5.7.1 The pressure transformation $v = u^{m-1}$

The theory of the porous medium equation has been much influenced by the consideration of the so-called pressure variable, which is $v = u^{m-1}$ with or without a normalizing constant factor. Using this transformation (without extra factor) on the FDE, we get the equation

$$v_t = v\Delta v - \kappa|\nabla v|^2, \qquad \kappa = 1/(1-m), \tag{5.39}$$

which applies for all $m \ne 1$. In the FDE range $m < 1$, v is an inverse power of u and the questions about vanishing of u transform into equivalent blow-up problems for v.

There are some features to be noted. The right-hand side of the pressure equation is quadratic in both terms, simpler to deal with in principle than the original equation, though this guess is not justified in practice. The former exponent m has become now part of a coefficient. The values $m = 1$, m_c and 0 correspond, respectively, to $\kappa = \infty$,

$n/2$ and 1, which does not seem very illuminating. On the other hand, the term $\kappa|\nabla v|^2$ works in the direction of inhibiting blow-up. We have seen above that there is no blow-up for $m > m_c$ which corresponds to $\kappa > n/2$. This chapter has shown that many solutions have blow-up when $\kappa < n/2$, and there is no blow-up if $\kappa = n/2$. We conclude that the existence of blow-up depends on a coefficient, which is a sign that we are moving in a borderline situation.

The questions we are discussing (does u become zero, where, why and in which way?) and their answers, translate into the following questions:

(i) Are there solutions v that become infinite after some finite time for an equation with given κ? which data correspond to blow-up solutions? (existence of blow-up).

(ii) Where and how does this happen? (mode of blow-up).

The second question will be discussed more in earnest in Chapter 7.

5.7.2 The Transformation $w = u^m$

Making use of this new transformation in one space dimension, we pass from the FDE to

$$w_t = w^{\alpha}\Delta w, \qquad \alpha = (m-1)/m. \tag{5.40}$$

Nothing really new seems to happen for $m > 0$, which means $\alpha < 1$, so that the diffusivity coefficient w^{α} is sublinear. Now $m = 1, m_c$ and 0 become $\alpha = 1, 2/(n-1)$ and ∞, respectively. On the other hand, for $m < 0$ we observe that w is an inverse power of u, so that the problems related to positivity (i.e., whether u vanishes or not), transform again into blow-up problems for w, i.e., whether w becomes infinity and how.

The above transformation does not apply in the limit case $m = 0$, but we may perform a suitable change, and then the case falls into a similar pattern. Just observe that the original equation is $u_t = \Delta(\log u)$, hence the transformation must be replaced by $w = -\log u$, i.e., $u = e^{-w}$. We get

$$w_t = e^{w}\Delta w, \tag{5.41}$$

with $u = 0$ transformed again into $w = +\infty$. Formally, this case is a limit of the previous one as $\alpha \to \infty$.

The same questions about blow-up can be put in this context.

5.8 Comments, extensions and historical notes

Section 5.1 The critical line $2p + mn = n$ we have introduced is relevant for the PME/FDE posed in the whole space; the presence of boundary conditions would completely alter the picture. Note that there are studies of extinction under different boundary conditions, see [DD79].

We recall that solutions with initial data that do not decay at infinity need not extinguish or even go to zero as $t \to \infty$. Thus, the solution with constant initial data is constant in space-time. Even solutions that decay slowly at infinity do not extinguish, e.g., the self-similar solutions with initial data $u_0(x) = c|x|^{-\gamma}$, $\gamma < (1-m)/2$.

On the other hand, solutions in L^p spaces with p smaller than critical may have a singular behaviour at say $x = 0$ for all times and not have extinction in finite time. An extreme case in that direction is the case of a Dirac delta $u_0(x) = M\,\delta(x)$. As Brezis and Friedman proved, the limit of any reasonable approximation is $u(x,t) = M\,\delta(x)$, so that no diffusion takes place at all (we ask the reader to prove a simple version of that statement in Exercise 5.1).

As we have said, the peculiar effects are not tied to non-uniqueness. We recall in this respect that some nonlinear evolution equations admit two distinct solutions corresponding to the same singular initial data. One of them belongs instantaneously to L^1, while the other remains singular. See [Na58], [BrC96] or [GV97]. This is not the case here.

Section 5.2 The first extinction result for L^p spaces is due to Bénilan and Crandall, [BC81]. Here is the result and the argument:

Proposition 5.18 *Let $0 < m < m_c$ and let u be a solution of the FDE with initial data in $L^q(\mathbf{R}^n)$, $q = (1-m)n/2$. Then the solution vanishes identically in a time that can be estimated above by $T \le c\|u_0\|_q^{1-m}$.*

Proof Suppose that the solution is non-negative and smooth, so that the calculations are justified. Then, for every $q > 1$ we have by the usual methods of integration by parts

$$\frac{d}{dt}\int_\Omega u^q\,dx = -\int_\Omega qmu^{m+q-3}|\nabla u|^2\,dx \le -C(m,n,q)\left(\int_\Omega u^p\right)^r$$

with $p = (m+q-1)n/(n-2)$ and $r = (n-2)/n$. We have used the Sobolev embedding in the last inequality. We now choose $q = n(1-m)/2$ and put $I_q = \int u^q\,dx = \|u(t)\|_q^q$ to get

$$\frac{dI_q}{dt} \le -C\,I_q^\gamma, \qquad \gamma = \frac{n-2}{n}. \tag{5.42}$$

Since $\gamma < 1$, explicit integration of this ODE for I_q leads to extinction in a finite time T depending only on m, n and $I_q(0)$. More precisely, we get as long as $u > 0$ and the calculation is justified

$$\int_{\mathbf{R}^n} u^q(x,t)\,dx \le \left(I_q(0)^{2/n} - \frac{2C}{n}t\right)^{n/2}. \tag{5.43}$$

This estimate is conserved when we make an approximation process. □

The use of the Marcinkiewicz space in formulating a sharp extinction result was first announced in [ChV02]. We discuss the extinction for the FDE in a bounded domain in the monograph [V06], Section 5.10.

The fact that a solution that vanishes identically continues to be zero for later times holds for problems in the whole space or for problems in a bounded domain with zero boundary conditions; it would not be true if we work in a bounded domain and non-trivial boundary conditions appear later on.

Universal rate These results are new. The magical number $k(m, n)$ expresses the minimum rate of dissipation of this class of solutions; it seems to appear by magic, but we will see another manifestation of such a magical number in Section 8.2.1 when discussing logarithmic diffusion (for $m = 0$, $n = 2$) and there is a beautiful explanation in terms of geometry. In that case the norm is in L^1 and the number is 4π. This can be obtained just as a formal limit of our universal estimate above.

A very interesting **Open problem** would be to investigate its deep meaning.

Universal estimates Note that all non-negative solutions of the FDE for $m < 1$ satisfy the pointwise estimate

$$u_t \le \frac{u}{(1-m)t}, \tag{5.44}$$

but such an estimate is of small use in extinction problems. It proves at least that a solution that vanishes at a point does not rise up later. Note that for $m > 1$ the estimate goes in the other direction, $u_t \ge u/(m-1)t$, and the qualitative conclusion is that a solution that is positive at a point x_0 and time t_0, will be positive at this point for all later times $t > t_0$; no extinction is thus possible.

The material in the rest of the subsections on extinction spaces, necessary conditions and control of T is new.

Section 5.4 The backward effect is new to our knowledge.

The lack of a smoothing effect below the critical line is exemplified by the self-similar solutions with $\gamma > n/p_*$ that have a singular point at $x = 0$ and continue to have that singularity for positive times. This standing singularity, also called non-removable singularity, is what allows them to live for infinite time.

An extension. Getting out of $L^1_{\mathrm{loc}}(\mathbf{R}^n)$ As an extension of the previous results, we can consider the existence of solutions when the data are not locally integrable and the relevant estimates. As we have said, the theory of the PME and the HE allows for initial data to be taken in $L^1_{\mathrm{loc}}(\mathbf{R}^n)$ under some growth conditions as $|x| \to \infty$, with no condition for $m < 1$ if the cases of non-existence are interpreted as instantaneous extinction.

We want to comment here on the possibility of non-locally integrable data. In the case of the PME and the HE there are no solutions having such type of data; they would simply blow up at $t = 0+$. Moreover, when the data are approximated by an increasing sequence u_{0n} of nice integrable data, the solutions u_n tend to infinity everywhere (instantaneous blow-up). See for instance [ChV02b].

Subsection 5.5.1 The phenomenon of outgoing flux at infinity will be analysed in great detail in Chapter 8 in the transition cases $n = 1, 0 < m \le 0$ and $n = 2$, $m = 0$, since in those cases it causes non-uniqueness of solutions of the Cauchy problem.

Subsection 5.5.3 We recall that studying fluids and other continuous media through the motion of particles is called the Lagrangian approach. See Appendix II for a reminder on that issue.

Subsection 5.5.4 The analysis of the limit of the source solutions as $m \to 0$ with $n = 2$ was performed in [VER96].

Section 5.6 The exponential decay for $m = m_c$ is described in Chapter 6 of [GV03]. The examples of unboundedness are new, but are based on the construction of [ChV02].

Section 5.7. The material of this section is basically taken from [Va05], where the interest was only in the case $n = 1$, $-1 < m \le 0$, to be discussed in Section 8.1.

General information about blow-up problems for reaction-diffusion equations can be obtained from many sources, like [BE89, GV02, S4-87]. Many examples of asymptotic behaviour for nonlinear heat equations are worked out in the text [GV03]. The range $-1 < m \le 0$ is equivalent to $\gamma \in (1/2, 1)$, not very illuminating.

The reader is referred to [Va04c] for a detailed discussion of the one-dimensional pressure equation for this FDE, even if the emphasis there is laid on free boundary solutions.

Exercises and open problems

Exercise 5.1 Approximations of the delta function.

(i) Let $\varphi \ge 0$ be a bounded and continuous function with $M = \int_{\mathbf{R}^n} \varphi(x)\, dx > 0$, and consider it as the initial data for a solution of the FDE with $0 < m < m_c$. Let $T > 0$ be the extinction time of that solution. Put $\theta = n(1-m) - 2 > 0$, and use the scaling transformation

$$u_k(x,t) = k^n u(kx, k^{-\theta} t)$$

to construct solutions u_k with the same initial mass and with extinction times $T_k = k^\theta T \to \infty$. Show that for every $t > 0$ we have $u(x,t) \to M\,\delta(x)$ as $k \to \infty$ (in the weak sense).

(ii) Take the ZKB solutions U_m with $m > m_c$ and let $m \to m_c$. Prove that

$$\lim_{m \to m_c} U_m(x,t;M) = M\,\delta(x)$$

in $Q = \mathbf{R}^n \times (0,\infty)$.

This exercise suggests that *the Dirac mass is not diffused* by the FDE with critical or subcritical exponent, so that a Dirac delta at $x = 0$ that does not change in time is the correct answer to the question of how the mass diffuses with time.

Exercise 5.2 Pseudo-Barenblatt solutions. In line with Remark (2) at the end of Section 5.2 on the resemblance with the source-type or ZKB solution, we may adapt these solutions to the range of parameters $m < m_c$ putting $C \ne 0$ instead of $C = 0$ in the original formula.

(i) Derive the formulas

$$U^{1-m} = \frac{T-t}{k|x|^2 + C(T-t)^{2\beta}},$$

where $\beta = +1/(2 - n(1-m)) < 0$.

(ii) Describe the solutions that are obtained when $C > 0$. Show that these solutions suffer no change in shape with respect to the case $m > m_c$, even if the time dependence changes. Notice that they have bounded profiles with fat, non-integrable tails, even if the algebraic formula is the same as in the FDE with $m > m_c$.

Note that the main difference is that they do have an infinite integral, hence, they do not have a Dirac delta as initial data. What is then the initial trace?

(iii) Show that in the case $C < 0$ the solutions are defined away from a central hole. Write the precise domain where the formula applies. Note that since the solution is infinite on the border of the hole, it serves as a universal barrier for the influence of any solution with data supported in the hole. Cf. Exercise 3.4.

Exercise 5.3 Compute the exact value of k in formula (5.6).

Exercise 5.4 (i) Show that $L^{p_*}(\mathbf{R}^n)$ is an adapted extinction space.

(ii) Show that it is not adapted from below in the sense that there are solutions with initial data u_{on} with L^{p_*} norm going to 1 such that the solutions have extinction times $T_n \to 0$. Equivalently, the initial norms may go to infinity and the extinction times converge to 1.

Hint for (ii): Use for instance a monotone sequence of bounded and compactly supported functions u_{on} converging to the initial function of formula (5.2) with $T = 1$.

Exercise 5.5 Dependence on m. Let u be a radially symmetric solution of the FDE with exponent m, with $0 \le u \le 1$ and let T be the extinction time. Under the assumption that $u_t \le 0$, conclude that the FDE with exponent $m' \in (0, m)$ and initial data u_0 generates a solution that has an extinction time $T' \le T$. Prove the monotone dependence of T on m, Proposition 5.10.

Exercise 5.6 Dependence of the solutions on dimension.

(i) Assume that u_0 is a radially symmetric, bounded and rearranged function that when considered as the initial data for the FDE in dimension n, produces a solution that vanishes identically in a finite time T. Show that the same will happen in dimension $n' > n$.

(ii) Note that the explicit extinction solution $U(x,t)$ has a formula where the power is independent of dimension. Looking at the constant, show that the extinction time decreases with increasing n. Generalize this property to other solutions.

Exercise 5.7 This exercise completes the study of the orbits of the self-similar system of Section 5.4.1. Analyse the behaviour of the rest of the orbits that originate from the asymptote $X = -\gamma$ in the phase plane in the same way as the backward profiles of Theorem 5.14, but go either towards $X \to -\infty$ or towards $X \to +\infty$. Show that they are solutions with changing sign in the first case, with blow-up at a finite distance in the second.

Exercise 5.8

(i) Show that solution (5.37) can be written in the form:

$$u(x,t) = e^{-n\beta t} f(x\, e^{-\beta t}), \tag{5.45}$$

where $\beta > 0$. This form is called exponential self-similarity or third type of self-similarity, cf. [V06].

(ii) Show that the equation for the profile is

$$\Delta f^{(n-2)/n} + \frac{(n-2)\beta}{n} \nabla \cdot (\xi f) = 0,$$

which coincides with equation $\Delta f^m + m\beta \nabla \cdot (\xi f) = 0$, for the standard Barenblatt profile. Show that we must have

$$f(\xi) = (a|\xi|^2 + k)^{-n/2},$$

where $a = \beta/n$ and k are arbitrary. This extends the Barenblatt profiles to the critical exponent $m = m_c$.

(iii) Find this solution as part of a more general search for explicit solutions, not necessarily self-similar as follows. Consider the pressure $v = u^{m-1}$ and try formulas of the form

$$v(x,t) = F(x) + G(t).$$

Show that they succeed for $m = m_c$ and then find solution (5.37).

(iv) Define the Lagrangian velocity as in Appendix II by the formula $\mathbf{v} = \nabla p$ with $p = u^{m-1}/(1-m)$ to find

$$p(x,t) = \frac{n}{2}(ax^2 + ke^{2nat}), \qquad \mathbf{v} = nax.$$

This shows a linear velocity-position relationship to be compared with formula (AII.6) for the ZKB when $m > m_c$ and with (5.32) for $m < m_c$. Show that this determines trajectories of the form

$$x = x(0)e^{nat},$$

as could be predicted from inspection of the formula for the solution.

(v) Take the limit $k = 0$ to find the singular profile $u(x,t) = C\,|x|^{-n}$ as a stationary solution of the FDE with critical exponent. This shows an example of limit solution that does not decay in time while it decays in space with the maximal power allowed for non-integrable solutions. Show that this solution is the limit as $m \downarrow m_c$ of the singular IPSS solutions (2.11) written (after introduction of a time delay) in the form

$$U = \left(\frac{c_1 + C}{|x|^2}\right)^{\frac{1}{1-m}}.$$

Note that $c_1 = 1/k_1 \to 0$ as $m \downarrow m_c$ and C is arbitrary and be taken as fixed. Show that it is also the limit of the extinction solutions (5.2) as $m \uparrow m_c$.

Exercise 5.9 Fill in the details of the construction of Example 5.3 above.

Open problem The technical problem we pose is: to determine the best space that is attracted by $L^1(\mathbf{R}^n)$, or at least a better space than $\mathcal{X} = \bigcup\{M^p(\mathbf{R}^n) : 1 < p < p_*\}$. Note that the range of exponents is sharp since we know that $\mathcal{X}$ cannot contain $M^{p_*}(\mathbf{R}^n)$.

Exercise 5.10 Show that the self-similar solutions with data (5.26), with $\gamma \geq n$ and $m_c > m > 0$, are extended continuous solutions and limits of approximations with locally bounded data.

Hint: Use the results of [ChV02].

Exercise 5.11 Investigate what happens with the solutions of the PME ($m \geq 1$) when we take initial data with 'critical exponent' $U_0(x) = A\,|x|^{2/(m-1)}$ and $A > 0$. Find the equivalent for $m = 1$ and for $m \in (m_c, 1)$.

Hint: For $m > m_c$ the solutions grow, [ChV02]. For $m > 1$ they even blow up in finite time, [Ar86].

Let us end this topic with a beautiful and curious situation.

Exercise 5.12 Show that for the precise value $\gamma = (n-2)/m$ we get the constant-in-time extended solution

$$U(x) = A\,|x|^{-(n-2)/m}, \tag{5.46}$$

valid for $A \geq 0$ and $m_c \geq m > 0$. Prove that it is not a solution in $Q = \mathbf{R}^n \times (0, T)$ for any $m > m_c$. Find the correct form of the solution in those cases.

Hint: Justify that when $m \leq m_c$ it is a solution also at $x = 0$ in the sense of the preceding exercise. Note that we include also the borderline case $m = m_c$. The negative part can be explained through the appearance of Delta functions in the right-hand side (in the sense of distributions). For $m \in (m_c, 1)$ we mean by solution the broadest definition, extended continuous solution.

Exercise 5.13

(i) Justify the calculation of the mass in Subsection 5.5.4.

(ii) Compute the size of the ball B_r that contains a given part of the total mass, $M_1 = a\,M$ with $0 < a < 1$, at a given time $t > 0$, and find the formula

$$\frac{1-m}{4m} r^2 = \left(\frac{4\pi t}{aM}\right)^{\frac{1-m}{m}} (1 - a^{\frac{1-m}{m}}).$$

Sketch the graphs for $m \approx 0$.

(iii) Examine the particle approach to see how particles move to infinity for $m > 0$ and in the limit $m = 0$.

Exercise 5.14

(i) Construct simple examples of non-extinction to prove Corollary 5.8 based on self-similarity or on formula (5.13).

(ii) Show that solutions with initial data in $L^p(\mathbf{R}^n)$, $p < p_*$, that are bounded at some time $t > 0$ must necessarily extinguish in finite time.

6

Improved analysis of the critical line. Delayed regularity

We continue here the analysis of the functional properties of the evolution semigroup generated by the FDE on the critical line, i.e., when $m < m_c$ and p assumes the critical value $p_* = n(1-m)/2$. The question we address here is boundedness, i.e., finding conditions on u_0 under which function $u(\cdot, t)$ is bounded for all $t > 0$. In that respect, Theorem 5.1 implies that data in $M^q(\mathbf{R}^n)$ generate bounded solutions for $t > 0$ when $q > p_*$. In view of the example of formula (5.2), we cannot ask for a similar result when data are taken in $M^{p_*}(\mathbf{R}^n)$.

However, the study of that space offers a clue to interesting results that we want to develop in this chapter. Indeed, we have seen that some solutions having an initial singularity in $M^{p_*}(\mathbf{R}^n)$ evolve for a time keeping the singularity, and eventually vanish in finite time. This can be seen as an extreme form of regularization after some delay.

6.1 The phenomenon of delayed regularity

In order to better understand the dynamics of such processes, it is advisable to pass to a more general setting, like some space of locally M^{p_*} functions. Note that we cannot ask for extinction in finite time if the space contains in particular the constants, since constant initial data produce constant solutions. We then pose the following problem: *to find optimal or nearly optimal spaces of initial data that produce bounded solutions.*

Actually, the investigation of this question has led the author to a very interesting phenomenon: there are solutions that become bounded only after a finite time. The functional space where this phenomenon will be studied is

$$X_* = M^{p_*}(\mathbf{R}^n) + L^\infty(\mathbf{R}^n) \subset M^{p_*}_{\text{loc}}(\mathbf{R}^n).$$

Before we state the result, we recall from the preliminaries (Subsection 1.1.4) the definition of the functional

$$\mathrm{N}_p(f) = \lim_{A\to\infty} \|(|f| - A)_+\|_{M^p} \qquad (6.1)$$

which applies when f is a measurable function such that $(|f| - A)_+ \in M^p(\mathbf{R}^n)$ for some $A > 0$, i.e., for functions in the space $M^p(\mathbf{R}^n) + L^\infty(\mathbf{R}^n)$. Note that the limit is zero for $f \in L^p(\mathbf{R}^n)$ but is not zero in the case of model example (1.14).

The following result contains the main quantitative estimates that support the phenomenon of delayed boundedness indicated above.

Theorem 6.1 *Consider the FDE* (2.1) *in the range* $m < m_c$ *and let* $p_* > 1$ *be the critical exponent. Let* $u \geq 0$ *be a solution with initial data* u_0 *in the space* $X = M^{p_*}(\mathbf{R}^n) + L^\infty(\mathbf{R}^n)$. *Then, there is a time* $T > 0$ *after which the solution is bounded and continuous. More precisely, there is a constant* $c = c(m, n) > 0$ *such that*

$$T \leq c(m,n)N^{1-m}, \tag{6.2}$$

where $N = \mathrm{N}_{p_*}(u_0)$. *Moreover, for all* $t \leq (0, T)$ *we have*

$$\mathrm{N}_{p_*}^{1-m}(u(t)) \leq \mathrm{N}_{p_*}^{1-m}(u_0) - k(m,n)\,t, \tag{6.3}$$

with $k(m, n) = 1/c(m, n)$.

The delay in the onset of regularity is not artificial. Actually, when the data belong to $M^{p_*}(\mathbf{R}^n)$ we have proved in Theorem 5.2 that the solution vanishes identically after a time T but it is not bounded before, and the self-similar example (5.2) is a good example of that. We are interested here in the case where the data have an extra bounded component that makes possible that the solution never vanishes identically (for instance, if the conditions of Lemma 5.5 are not met), but still it becomes bounded after a finite time.

We recall that once the solution is shown to be bounded, known regularity theory implies that a locally bounded and non-negative solution of the FDE is continuous in some Hölder class, cf. [DiB93]. Moreover, when $m \leq 1$ it is even C^∞ smooth and positive, as long as it does not vanish identically. Therefore, we can call this phenomenon *delayed regularity*. This is a novel feature in our qualitative description of nonlinear diffusion equations. We want to propose for such phenomenon the name of *blow-down*, since it is in some sense the opposite of the well-known phenomenon of blow-up discussed in Section 5.7.

6.1.1 Preparation for the proof of Theorem 6.1

This can be reduced to proving some special situation by using some of the functional analytic tricks we have already seen, like the transformations based on the scaling properties of the equation, plus symmetrization and worst-case strategies.

1. Thus, we start by noting that by the assumption on the initial data of Theorem 6.1, $(u_0 - A)_+$ belongs to M^{p_*} for some $A > 0$, hence for all large enough A; moreover, the M^{p_*} norm is non-increasing as A grows.

2. We need the scaling group. For real parameters K, L, S the rescaled function

$$\tilde{u}(x,t) = K\,u(Lx, St) \tag{6.4}$$

is also a solution if the parameters are tied by the relation $K^{1-m}S = L^2$. This general transformation has interesting properties for functions in M^{p_*}. In fact, if we specialize to the case $S = 1$, so that we only do rescaling in u and x of the form

$$\tilde{u}(x,t) = K\, u(K^{(1-m)/2}x, t),$$

it is easy to see that the M^{p_*} norm is conserved, $\|\tilde{u}(t)\|_{M^{p_*}} = \|u(t)\|_{M^{p_*}}$.

As another example of the use of the transformation, we may use it with $L = 1$ to absorb the norm of the data. Thus, if $M = \|(u_0 - A)_+\|_{M^{p_*}}$, we choose the values $K = 1/M$ and $S = M^{1-m}$. In this way, since

$$(\tilde{u}_0 - AK)_+ = K(u_0(x) - A)_+ ,$$

it has M^{p_*}-norm $= 1$. In the limit $A \to \infty$, a similar relation holds for the functional N_p. Suppose that the theorem is proved for the latter class of solutions with $N_p(u_0) \le 1$ and the optimal estimate of the time is $c(m,n)$. Since we have

$$Ku(tS) = \tilde{u}(t),$$

the regularity time for a solution u with $N_p(u_0) = N$ would then be bounded above $T = c/S = cN_p^{1-m}$, as stated in (6.2).

3. On the other hand, we may use the symmetrization results of [Va04b], see Section 1.3, to prove the result only for a rearranged function. Note that we have to change the horizontal axis to the level $u = A$ (since the result applies to integrable functions) but this is no major problem, we just write the equation for $v = u - A$ which turns out to be a filtration equation for which symmetrization and mass comparison have been shown to hold.

Finally, by using the worst case strategy, we may further reduce the examination of the worst case, which is a function of the form

$$u_0(x) = C\,|x|^{-2/(1-m)} + A. \tag{6.5}$$

Therefore, we will concentrate on estimating the onset of regularity for that particular example.

6.1.2 Main lemmas

We are now ready to state and prove the basic quantitative estimate, on which the proof is built.

Lemma 6.2 *Let $(u_0 - 1)_+ \in M^{p_*}(\mathbf{R}^n)$ with norm 1. Then, for any $\varepsilon > 0$ small there exist constants $\tau > 0$, d, $\lambda \in (0,1)$ depending only on n, m, and $k_1, k_2 > 0$ depending also on ε, such that for all $0 < t \le \tau$*

$$\|(u(t) - (k_1 + k_2 t))_+\|_{M^{p_*}} \le (1+\varepsilon)\left(1 - \lambda\frac{t}{\tau}\right)^{1/(1-m)}. \tag{6.6}$$

For $m \ge 0$ we may take $\varepsilon = 0$. Moreover, the pointwise inequality (6.9) holds.

Proof (i) We treat first the case $m = 0$ (with $n \geq 3$) where the computations are less involved. The proof relies on finding a supersolution for the equation $u_t = \Delta \log(u)$, i.e., a function $U(x,t)$ such that $U_t \geq \Delta \log(U)$ in $\mathcal{D}'(Q_T)$. We try

$$U = d\left(\frac{T-t}{|x|^2} + K(t)\right) \tag{6.7}$$

with a suitable choice of constant $d > 0$ and increasing function $K(t)$ with $d\,K(0) = 1$. $T > 0$ is arbitrary at this stage. Note that $U(\cdot, t) \in L^1_{\mathrm{loc}}(\mathbf{R}^n)$ uniformly in $t \geq 0$ as long as $K(t)$ is bounded. Putting $\Omega = (T-t) + K(t)|x|^2$ and $|x| = r$, the supersolution condition inside the domain reads for $r > 0$

$$-dr^{-2} + dK'(t) \geq r^{1-n}\partial_r\big(-2(T-t)r^{n-2}\,\Omega^{-1}\big).$$

Since $\partial_r\Omega = 2Kr$, the right-hand side gives

$$-2(n-2)(T-t)r^{-2}\Omega^{-1} + 4K(T-t)\Omega^{-2}.$$

Substituting and multiplying everything by $r^2\Omega^2$, we get the necessary condition

$$dK'(t)r^2\Omega^2 + 2(n-2)(T-t)\Omega \geq d\Omega^2 + 4K(T-t)r^2.$$

Putting $r = y(T-t)^{1/2}$ we get $\Omega = (1+Ky^2)(T-t)$, and the condition reads

$$dK'(t)(T-t)y^2(1+Ky^2)^2 + 2(n-2)(1+Ky^2) \geq d(1+Ky^2)^2 + 4Ky^2.$$

First of all, d has to be equal or smaller than $2(n-2)$ for the inequality to be true for y small, and then $K'(t)(T-t)$ has to be large enough. Note that $c_m = 2(n-2)$ for $m = 0$ in formula (5.2).

(ii) Now that we have a supersolution in Q_T, we choose T so as to make the M^{p_*}-norm of $U_0(x) - 1 = dT\,|x|^{-2}$, equal to one. This means that $\kappa_{p_*}dT = 1$. We can now make a comparison of concentrations between the original solution u and U and conclude that $u(t)$ will be less concentrated than $U(t)$. We notice that for any $t \in (0,T)$ the norm of $(U - K(t))_+$ is

$$(1 - (t/T)),$$

and this is smaller than 1. By comparison, the result follows with $\tau \in (0,T)$, putting $\lambda = \tau/T$ and $k = \max\{d\,K'(t) : 0 < t \leq \tau\}$. This ends the proof for $m = 0$.

(iii) Let us address the case $m \neq 0$. We take a supersolution of the form

$$U = d\left(\frac{T-t}{|x|^2} + K(t)\right)^{1/(1-m)}. \tag{6.8}$$

We put again $|x| = r = y(T-t)^{1/2}$ and $\Omega = (T-t) + K(t)|x|^2 = (T-t)(1+Ky^2)$. The supersolution condition reads

$$\frac{d}{1-m}\Omega^{m/(1-m)}r^{-2/(1-m)}(K'r^2-1) \geq -\frac{2d^m}{1-m}(T-t)r^{1-n}\partial_r(r^{n-2-\frac{2m}{1-m}}\Omega^{\frac{2m-1}{1-m}}).$$

The derivative of the term $r^{1-n}\partial_r(\cdots)$ in the right-hand side equals

$$k_m r^{-\frac{2}{1-m}}\Omega^{\frac{2m-1}{1-m}} + \frac{2(2m-1)}{(1-m)}Kr^{-\frac{2m}{1-m}}\Omega^{\frac{3m-2}{1-m}},$$

with $k_m = (n(1-m)-2)/(1-m) > 0$. Substituting and multiplying by the quantity $\Omega^{-(3m-2)/(1-m)}r^{2/(1-m)}$, we get

$$d\,\Omega^2(K'r^2-1) \geq -2k_m(T-t)\Omega + \frac{4(1-2m)}{(1-m)}d^m K(T-t)r^2,$$

or

$$\begin{aligned} d\,(T-t)K'\,y^2(1+Ky^2)^{2)} + 2k_m d^m(1+Ky^2) \\ \geq d\,(1+Ky^2)^2 + \frac{4(1-2m)}{(1-m)}d^m Ky^2. \end{aligned}$$

We again see that $d \leq c_m$ guarantees that the inequality is satisfied for small y, and then $K'(t)(T-t)$ has to be big enough to complete the inequality.

Now that we have a supersolution, we make comparison in the same way as at in the end of step (ii). The details change a bit. For the initial condition one needs $u(x,0) \leq U(x,0)$, that is implied by

$$\left(\frac{M}{\kappa|x|^{2/(1-m)}}+1\right)^{1-m} \leq d^{1-m}\left(\frac{T}{|x|^2}+K(0)\right).$$

If $m \geq 0$, $1-m \leq 1$, this holds with $d^{1-m}K(0) = 0$ and any $T \geq (M/d\kappa)^{1-m}$ and we conclude as before.

However, if $m < 0$ we need to use the inequality $(X+Y)^p \leq (1+\varepsilon)X^p + c_\varepsilon Y^p$ which is true for $X, Y \geq 0$ and $p = 1-m > 1$ for any given $\varepsilon > 0$ if c_ε has at least a value larger than 1 that depends only on ε and m. Applying this to the left-hand side, we see that the condition holds if

$$T \geq (1+\varepsilon)(M/d\kappa)^{1-m}, \qquad d^{1-m}K(0) = 1 + c_\varepsilon.$$

We will select $T \geq (1+\varepsilon)(M/d\kappa)^{1-m}$ independent of ε and $M = 1$. With these choices, after a time $0 < t < T$ we get $u(x,t) \leq U(x,t)$, hence

$$u(x,t) \leq d\left(\frac{T-t}{|x|^2}+K(t)\right)^{1/(1-m)} \leq (1+\varepsilon)d\left(\frac{T-t}{|x|^2}\right)^{1/(1-m)} + c_\varepsilon d K(t)^{1/(1-m)}.$$

We restrict t to the interval $(0, \tau)$ with $\tau = T/2$ to get $K'(t) \le k_2$, hence $K(t) \le 1 + \varepsilon + k_2 t$ and we get in all cases for $0 < t < \tau$

$$u(x,t) \le \frac{1}{\kappa}\left((1+\varepsilon)\frac{1-(t/T)}{|x|^2}\right)^{1/(1-m)} + (dc_\varepsilon + C_1\, t)^{1/(1-m)}. \tag{6.9}$$

The result follows since κ is the factor that comes in the M^{p_*} norm of this form of function (the ε is slightly changed). □

Using the scaling transformation allows us to generalize this result.

Lemma 6.3 *Let $(u_0 - A)_+ \in M^{p_*}(\mathbf{R}^n)$ with norm 1 for some $A > 0$. For all $0 < t \le \tau$ we have*

$$\|(u(t) - A(k_1 + k_2 t))_+\|_{M^p} \le (1+\varepsilon)\left(1 - \lambda\frac{t}{\tau}\right)^{1/(1-m)} \tag{6.10}$$

with τ, λ and k as before.

Proof We perform the scaling transformation with $S = 1$ and $K = 1/A$, so that $L = A^{-(1-m)/2}$. Then

$$(u_0(x) - A)_+ = A(\tilde{u}_0(A^{(1-m)/2}x) - 1)_+$$

and the Marcinkiewicz norm is the same. The result follows from Lemma 6.2 since

$$(u(x,t) - cA)_+ = A(\tilde{u}(A^{(1-m)/2}x, t) - c)_+$$

with $c = 1 + kt$. □

A second use of the scaling transformation allows us to treat norms different from 1.

Lemma 6.4 *Let $(u_0 - A)_+ \in M^p(\mathbf{R}^n)$ with norm M for some $A > 0$. Then, for all $0 < t \le M^{1-m}\tau$ we have*

$$\|(u(t) - A(k_1 + k_2 M^{m-1} t))_+\|_{M^p} \le (1+\varepsilon)M\left(1 - \frac{\lambda t}{M^{1-m}\tau}\right)^{1/(1-m)}. \tag{6.11}$$

with τ, λ and k as before.

Proof We use the transform now with $L = 1$ so as to absorb the norm. We put $K = 1/M$ and $S = M^{1-m}$, hence $u(x,t) = M\tilde{u}(x, M^{m-1}t)$. In this way

$$(\tilde{u}_0 - AK)_+ = K(u_0(x) - A)_+$$

has norm 1. Lemma 6.3 implies that for $0 < t' < \tau$ with $t' = tM^{m-1}$ we have

$$\|(\tilde{u}(t) - AK(1 + kt'))_+\|_{M^p} \le \left(1 - \lambda\frac{t'}{\tau}\right)^{1/(1-m)},$$

hence the result. □

Corollary 6.5 *There exists a constant* $a(m,n) > 0$ *such that*

$$\frac{d}{dt}\mathrm{N}_*(u(t))^{1-m} \le -a. \tag{6.12}$$

Proof We only need to pass to the limit $A \to \infty$ in the previous result. □

We need another estimate.

Lemma 6.6 *If* u_0 *satisfies*

$$u_0^m(x) \le A|x|^{2m/(1-m)} + C, \tag{6.13}$$

the same happens for all $t > 0$.

The proof is just checking that the right-hand side is a supersolution of the equation.

6.1.3 End of proof of Theorem 6.1

(i) We may assume that the initial function is rearranged, as explained above. We will prove that, after a finite time, the solution enters a better space, $L^q_{\rm loc}(\mathbf{R}^n)$ with $q > p$. This will done by an iteration process, slowly improving the local Marcinkiewicz norm.

We take an initial function u_0 such that $(u_0 - A)_+$ has norm M. Using Lemma 6.4, after a time $\tau_1 = \tau M^{1-m}$ we get a function $u(t)$ such that $(u(\tau_1) - cA)_+$ has norm equal or less than $M\lambda$.

We now shift the origin of time to τ_1 and apply Lemma 6.4 again. Put $\epsilon = \lambda^{1-m} < 1$. It follows that there exists a time τ_2 such that

$$\tau_2 - \tau_1 = \tau_1\epsilon$$

and

$$\|(u(\tau_2) - c^2A)_+\|_{M^p} \le M\lambda^2.$$

We have therefore a sequence of times τ_k, $k = 1, 2, \ldots$, converging to a time $\tau_* = \tau_1/(1-\epsilon)$. By virtue of the last lemma of the previous section we know that for $t \ge \tau_k$ we have

$$(u(t) - c^kA)_+ \le CM\lambda^k|x|^{-2/(1-m)}.$$

The whole set of inequalities holds therefore at $t = \tau_*$.

We contend that $u(\tau_*)$ belongs to a better L^p space. Indeed, let us calculate the integral $\int_B u(x)^q\,dx$ in the ball B of radius 1 with $q = p_* + \delta$ for some small $\delta > 0$ to be determined. We divide the ball into annuli $\mathcal{A}_k = \{r_{k+1} \le |x| \le r_k\}$ with $r_k = (\lambda/c)^{k(1-m)/2}$. In this way we ensure that

$$u(\tau_*) \le KM\lambda^k|x|^{-2/(1-m)}$$

in $\mathcal{A}_k$ for some K. Then,

$$\int_{\mathcal{A}_k} u(\tau_*)^q\,dx \le \sum_k KM^q\lambda^{kq}\int_{\mathcal{A}_k}|x|^{-2q/(1-m)}dx \le K_1\sum_k \lambda^{kq}r_k^{n-2q/(1-m)}.$$

Since $q = p_* + \delta$, and $p_* = n(1-m)/2$, we get

$$\int_{A_k} u(\tau_*)^q \, dx \le K_2 \sum_k \lambda^{kq} (c/\lambda)^{k\delta} = K_2 \sum_k \lambda^{kp} c^{k\delta}.$$

Taking into account that $\lambda < 1$, we only need to put δ small so that $c^\delta < (1/\lambda)^{p_*}$ for this integral to converge. This means that $u(\tau_*) \in L^q_{\rm loc}(\mathbf{R}^n)$.

(ii) The proof that u is bounded after $t = \tau$ is now easy. Let us shift the origin of time to τ_* so that we now assume that $u_0 \in M^q_{\rm loc}(\mathbf{R}^n)$. We know, by Theorem 5.1, that there exists a self-similar solution u_q with data

$$u_0(x) = \frac{C}{|x|^{n/q}}$$

for all $q > p_* = n(1-m)/2$. For large C, u is less concentrated than u_q at $t = 0$. We also know that for any $|x| = R > 0$ the value of u is bounded for all times $t > 0$. For large C we can check that the value of $u_q(x,t) \ge u(x,t)$ for $|x| = R$ and $0 < t < 2\tau$ for some small τ. By comparison, u is bounded and the result follows. □

6.2 Immediate boundedness

As an almost immediate consequence of this theorem, we obtain a criterion for boundedness for all $t > 0$.

Theorem 6.7 *Let $m < m_c = (n-2)/n$. Let u be solution to the FDE in $\mathbf{R}^n$ with initial data $u_0 \in L^{p_*}(\mathbf{R}^n) + L^\infty(\mathbf{R}^n)$. Then, $u(t)$ is locally bounded (and smooth) for all $t > 0$. In fact, the result holds for $u_0 \in M^{p_*}(\mathbf{R}^n) + L^\infty(\mathbf{R}^n)$ with $\mathrm{N}(u_0) = 0$.*

The proof of this result follows from the previous theorem by observing that in that case $\mathrm{N}(u_0) = 0$, hence $T = 0$. □

Note that exponents $m \le 0$ are also covered. The results are of course true for $u_0 \in L^p(\mathbf{R}^n) + L^\infty(\mathbf{R}^n)$ if $p > p_*$, but they are false $p < p_*$, as shown by the explicit backward self-similar solutions constructed in Section 5.4.1.

We now give an example to show that immediate boundedness is not to be expected for most M^{p_*} data. Actually, the explicit solution (5.2) shows that solutions can have a positive extinction time with a singularity for all previous times. Comparison says that when $m < m_c$ and $u_0 \in L^1_{\rm loc}(\mathbf{R}^n)$ with

$$u_0(x) \ge A\,|x|^{-2/(1-m)},$$

then the singularity at $x = 0$ persists for a time $0 < t < T$ with

$$T = a(m,n)\,A^{1-m}. \tag{6.14}$$

In comparison, Theorem 6.1 says that when

$$u_0(x) \le A\,|x|^{-2/(1-m)},$$

then the singularity at $x = 0$ stops at least after a time

$$T = b(m,n)\,A^{1-m}.$$

Both results imply that there is a singularity for a time that we would like to know how to determine.

Two open problems

(1) Is the condition $\mathrm{N}(u_0) = 0$ if and only if for immediate boundedness inside the class $u_0 \in M^{p_*}(\mathbf{R}^n) + L^\infty(\mathbf{R}^n)$?

(2) Calculate rates of decay of $\|u(t)\|_\infty$ for $u_0 \in L^p(\mathbf{R}^n)$.

6.3 Comments and historical notes

Section 6.1 The description of the phenomenon of delayed regularity is new; I thank Profs. L. Boccardo and A. Pozio for conversations on this issue. We will discuss further the existence of solutions with delayed regularity in dimension $n = 1$ for exponents $m < -1$ in Theorem 10.3, in the course of the study of existence of source solutions with a background. The method used in that case is completely different from the technique of this chapter.

We have studied a related phenomenon of blow-down in $n = 2$ and $m = 0$ in [V05b]. There the name blow-down is proposed.

***p*-Laplacian equations** The L^p–L^∞ smoothing effect on the extinction line has been shown to be true for the p-Laplacian equation in [HV81]. The proof is quite simple, but it does not adapt to the FDE.

Exercises

Exercise 6.1 Use Theorem 6.7 to improve the result on dependence of the extinction time on the L^1-norm of the data stated in Theorem 5.9 to classes of data that are bounded in $L^{p_*}(\mathbf{R}^n)$.

Exercise 6.2 What are the extinction time and the behaviour of the solution with data

$$u_0(x) = |x|^{-2/(1-m)}\chi(B_R)?$$

Hint: Use scaling in u-x to see that R plays no role in the extinction time.

7

Extinction rates and asymptotics for $0 < m < m_c$

The chapter deals with the actual asymptotic behaviour of the solutions of the FDE in the exponent range $0 < m < m_c$. This behaviour depends on the class of initial data. We are interested in 'small solutions' that extinguish in finite time, according to the results of the Chapter 5.

We will concentrate on solutions that start with initial data in $L^1(\mathbf{R}^n)$, or solutions that fall into this class for positive times prior to extinction. In the range $m > m_c$ the ZKB solutions provide the clue to the asymptotics for all non-negative solutions with L^1-data. We already know that these solutions do not exist (at least as finite mass solutions) in the range $m < m_c$. So we need to look for an alternative. This alternative has to take into account the fact that solutions with M^{p_*}-data disappear (vanish identically) in finite time, cf. Theorem 5.15. The class of self-similarity solutions of Type II will provide the clue to the asymptotic behaviour near extinction. Therefore, we start the chapter by studying that type of functions.

We identify in Section 7.2 the solutions that play the main role in the present setting. They are a special class of self-similar solutions with fast decay at infinity, that represent an example of self-similarity with anomalous exponents, also called by Ya. B. Zel'dovich *self-similarity of the second kind.* Anomalous means that the exponents cannot be obtained from dimensional considerations, as we have done in the ZKB case by means of the scaling group. This kind of solution is of great interest for its analytical difficulty, and they are found in many applications. We calculate the anomalous exponents of those solutions following the work of J. King, M. Peletier and H. Zhang, and identify them as minimal extinction exponents in a precise sense. The study is complemented in Section 7.3 with the investigation of admissible extinction rates.

We then pass to the question of convergence of general classes of solutions towards the KPZ solutions, Section 7.4. This is in analogy to the results for the KZB profiles for $m > m_c$. The results are more complete when $m = m_s = (n-2)/(n+2)$, where the problem has a nice geometrical interpretation in terms of the famous Yamabe problem of Riemannian geometry, Section 7.5. The convergence results are restricted to radial solutions when $m \neq m_s$. In those matters, we expand on basic results of del Pino-Saez, and Galaktionov and King using the tools provided by the previous two chapters.

We end the chapter with a review of what is known for the Dirichlet problem in a ball, Section 7.6. This is a complement for the reader's convenience.

7.1 Self-similarity of Type II and extinction

We provide in two subsections the general facts about this type of solutions.

7.1.1 Self-similarity and elliptic equations

Apart from the self-similar forms studied in Section 3.2, there is a second type of similarity that corresponds to the Ansatz

$$U(x,t) = (T-t)^{\alpha} f(x\,(T-t)^{\beta}), \tag{7.1}$$

to be valid for $x \in \mathbf{R}^n$ and $t < T$. This is called *Type-II self-similarity*, as opposed to the Type-I self-similarity Section 3.2. It is sometimes called *backward self-similarity* because solutions of the new type do not extend forward in time past T, but they extend backwards in time as far as $t \to -\infty$. In this situation, the similarity exponents are related by $\alpha(m-1)+2\beta = -1$, i.e.,[1]

$$\alpha(1-m) = 2\beta + 1, \tag{7.2}$$

and the equation for the profile $f = f(\eta)$ is

$$\Delta f^m + m\alpha f + m\beta\,(\eta \cdot \nabla f) = 0 \tag{7.3}$$

for $\eta \in \mathbf{R}^n$, formally the same as in Type I. This is a nonlinear elliptic PDE that is best written in terms of $g = f^m \ge 0$ as

$$-\Delta g = a\, g^p + b\,\eta \cdot \nabla g^p, \quad p = 1/m. \tag{7.4}$$

where $a = m\alpha$, $b = m\beta$. The case $\beta = 0$ (no scaling in space; in other words, separation of variables) is important because it becomes the well-known Emden–Fowler equation

$$-\Delta g = c g^p, \qquad p > 1. \tag{7.5}$$

To end this introduction, let us remark that the use of Type-II self-similarity is not restricted to the range that focuses our attention at this moment, $m > 0$. Indeed, the consideration of ranges $m < 0$ will be important for later chapters, and the Emden–Fowler equation has the non-standard exponent $p < 0$, already considered by Joseph and Lundgren in their fundamental work [JL73]. The case $m = 0$ is also included, but it is special: then we must take $g = \log(f)$ and the equation has an exponential nonlinearity: $-\Delta g = e^g$.

[1]Note that the relation was $\alpha(m-1)+2\beta = 1$ for Type I. The reader is warned that the signs of α and β in formula (7.1) are sometimes changed in the literature. Such changes affect in particular the formula relating them.

Renormalized flows Even for solutions that are not self-similar, the previous ideas are useful, since they lead us to define the renormalized flow. This concept will be quite useful in the analysis of solutions that vanish in finite time when analysing the times close to the extinction time (asymptotic behaviour).

(R1) The basic idea is the following: we take $T > 0$ and replace the variables u and t by

$$v(x,\tau) = (T-t)^{-1/(1-m)}u(x,t), \qquad \tau = -\log(T-t). \tag{7.6}$$

Then, the FDE $u_t = \Delta(u^m/m)$ transforms into the evolution equation

$$v_\tau = \Delta(v^m/m) + \frac{1}{1-m}v. \tag{7.7}$$

Note that the change from u to v is highly sensitive of the value of T used in the transformation: if T is chosen as the extinction time of u, then v is a global solution of (7.7), i.e., it exists and is nonzero for $\tau_0 < \tau < \infty$. However, when T is smaller or larger than $T(u_0)$, then v either blows up in finite time of goes to zero. Studying the possible stabilization of $v(\tau)$ (in the case where T is the extinction time) towards a stationary solution of

$$\Delta v^m + \frac{m}{1-m}v = 0 \tag{7.8}$$

is a key tool in the asymptotic theory.

(R2) This is not the only possibility we will use. When we want to study situations where non-self-similar solutions approach a self-similar solution of the form (7.1) with $\beta \neq 0$, then the previous renormalization is replaced by

$$v(\eta,\tau) = (T-t)^{-\alpha}u(x,t), \qquad \tau = -\log(T-t), \qquad x = \eta\,(T-t)^{-\beta}, \tag{7.9}$$

which leads to the equation

$$v_\tau = \Delta_\eta(v^m/m) + \alpha\,v + \beta\eta\cdot\nabla_\eta v. \tag{7.10}$$

The stationary states of this equation are just (7.3), as we wished.

7.1.2 *ODE analysis of radial profiles*

Let us now concentrate on radially symmetric profiles $f = f(\xi)$, $\xi = |\eta| > 0$. As in Section 3.3.2, the ODE for the profile f is

$$\xi^{1-n}(\xi^{n-1}f(\xi)^{m-1}f'(\xi))' + \beta\xi\,f'(\xi) + \alpha\,f(\xi) = 0. \tag{7.11}$$

The difference is only noticed in the fact that when we write $\alpha = \gamma\beta$, then

$$\beta = \frac{1}{\gamma(1-m)-2}, \qquad \alpha = \frac{\gamma}{\gamma(1-m)-2}, \tag{7.12}$$

which is just the opposite of the formulas (3.15) for type-I similarity. Proceeding just as there, we introduce new variables

$$X = \frac{\xi f'(\xi)}{f(\xi)}, \qquad Z = \xi^2 f^{1-m}, \tag{7.13}$$

and $r = \log \xi$. Then, we get the system

$$\begin{cases} \dot{X} = (2-n)X - mX^2 - \beta(\gamma + X)Z, \\ \dot{Z} = (2 + (1-m)X)Z, \end{cases} \tag{7.14}$$

where $\dot{X} = dX/dr$, $\dot{Z} = dZ/dr$ and β is given in terms of γ by (7.12). With respect to system (3.9) with the same value of γ, there is only a difference, the sign of β as a function of γ and the variable Z instead of Y.

The use of new variable Z is made on purpose to signify that difference. Moreover, there is a way of reducing the study of both types of similarity to the same phase-plane. Let us explain this last detail. Indeed, if $(X(r), Z(r))$ is an orbit of system (7.14), and we put $X_1(r) = X(r)$ and $Y_1(r) = -Z(r)$, then $(X_1(r), Y_1(r))$ is an orbit of system (3.9) with the value of β in terms of γ given by (3.15), as corresponds to Type-I self-similarity (i.e., $\beta_1(\gamma) = -\beta(\gamma)$, and also $\alpha_1(\gamma) = -\alpha(\gamma)$). Moreover, X and X_1 travel in the same direction, though Z and Y_1 are reflections of each other (with respect to the X-axis). Summing up, a symmetry around the X-axis allows us to pass from the study of the first system to the study of the second. We use the convention that the upper half-plane of System (3.9) represents Type-I self-similarity, while the lower half-plane of the same system portrays the orbits of Type-II self-similarity.[2]

7.2 Special solutions with anomalous exponents

We now discuss the existence and properties of a special family of self-similar solutions that play a prominent role in understanding the phenomenon of extinction. Actually, these solutions describe the asymptotic behaviour near extinction of large classes of solutions. The solutions have the self-similar form

$$\overline{U}(x,t) = (T-t)^{\alpha} F(x\,(T-t)^{\beta}; a) \tag{7.15}$$

for a special value of β, let us call it $\overline{\beta}$, and $\overline{\alpha}(1-m) = 2\overline{\beta} + 1$; a is a free parameter. The solutions are characterized by having fast decay as $x \to \infty$.

The existence of such a family of solutions has been discussed by King in his seminal paper [Ki93], where the main asymptotic results on fast diffusion were examined and surveyed by formal methods. A rigorous analysis was then performed by M. Peletier and Zhang [PZ95], who showed the existence of the relevant asymptotic profiles as a

[2] We warn the reader that the use of the lower phase plane can easily lead to confusion in assigning the signs of the different parameters and variables. We introduce this variant here because it appears quite often in the literature on self-similarity.

consequence of a careful phase-plane analysis as proposed at the end of the last section. We will expand here on these results.

Let us recall the main results of paper [PZ95] stated in our notation.

Theorem 7.1 *For every $0 < m < m_c$ there is a unique $\overline{\gamma} = \overline{\gamma}(m)$ and a unique self-similar solution of the FDE of the form*

$$u(x,t) = (T-t)^{\overline{\alpha}} F(x(T-t)^{\overline{\beta}}) \tag{7.16}$$

(up to scale transformation) with $\overline{\alpha}$ and $\overline{\beta}$ given in terms of $\overline{\gamma}$ by (7.12), *and such that F is bounded and has a spatial decay as $|x| \to \infty$ of the form*

$$F(\xi) \sim C\,\xi^{-(n-2)/m}. \tag{7.17}$$

Moreover, $\overline{\alpha}$ and $\overline{\beta}$ depend continuously on m and $\overline{\alpha} \to \infty$ as $m \to m_c$, while it tends to zero as $m \to 0$.

By unique up to scale transformation we mean that the special solution is shown to be unique apart from a scaling due to the invariance of the equation for F under the transformation:

$$F(\eta; a) = a^{-2/(1-m)} F(\eta/a; 1). \tag{7.18}$$

We may normalize the family by fixing $F(0;1) = 1$. This is analogous to the free parameter of the ZKB solutions.

We will call the profiles f the KPZ profiles from the names of the authors. Exponent $\overline{\beta}$ (or $\overline{\alpha}$) is not determined (in terms of m and n) by simple algebra as we have often done before, but as a result of a different and more complicated computation that we want to examine here: in fact, it is the result of the analysis of the ordinary differential equation satisfied by function F; in phase plane terms, it is the value for which a certain connection exists. It is therefore a kind on nonlinear eigenvalue. Such a value is called an *anomalous exponent* in the literature on self-similarity, cf. [Ba79] and it is also referred to as *similarity of the second kind*. We will avoid using that term here since it can be confused with Type-II similarity.

Let us now add some comments:

(i) The decay rate (7.17) is called fast, in comparison to the more usually found case $f(\xi) \sim \xi^{-2/(1-m)}$ of the separate-variable solution (5.2) of Chapter 5, which will be called slow. As for time decay, only in one case it coincides with the one already found in Chapter 5, $\alpha = 1/(1-m)$.

(ii) The variation of $\overline{\beta}$ and $\overline{\alpha}$ with m has a direct quantitative interest, since the latter indicates the actual decay rate of the L^∞ norm of the solution. The former has a less obvious meaning, but it is also important since it is related to the way solutions spread.

(iii) Special attention will be paid below to the case when $m = (n-2)/(n+2)$ (that we will call henceforth the Sobolev exponent, m_s, since it is the inverse of the Sobolev exponent of the elliptic theory). Then, it can be easily checked that, as

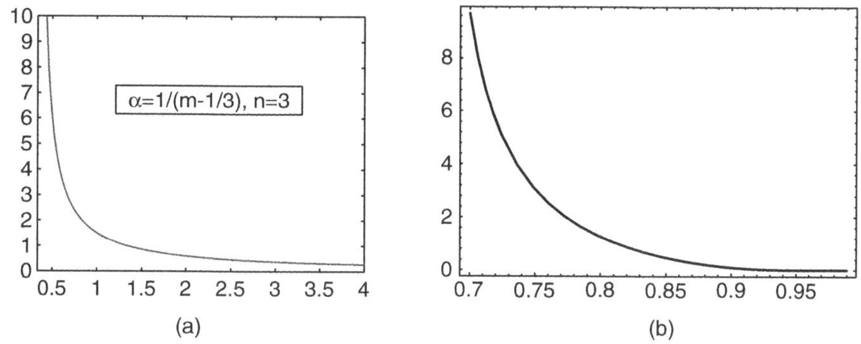

Figure 7.1 ZKB and KPZ exponents for $n = 3$. Graph (b) plots α versus $1 - m$.

already explained in [Pe81], $\overline{\beta} = 0$, $\overline{\alpha} = 1/(1-m) = (n+2)/4$, the self-similar solution is in fact a separate variables solution, and the profile F can be found explicitly

$$F(x; a) = \left(a + \frac{|x|^2}{4na}\right)^{-(n+2)/2} \tag{7.19}$$

where $a > 0$ is arbitrary. Note that the special solution is then

$$U(x,t) = (T-t)^{\frac{n+2}{4}} F(x), \tag{7.20}$$

to be compared to the singular solution (5.2).[3] The equation with Sobolev exponent plays a role in differential geometry, and will be studied in more detail in Section 7.5.

We continue with the statement of some relevant properties of the anomalous exponents.

Theorem 7.2 *Both $\overline{\alpha}$ and $\overline{\beta}$ are monotone increasing, analytic functions of m. We have $\overline{\beta} = 0$ precisely for $m = m_s := (n-2)/(n+2)$. Moreover, $\overline{\alpha} \to 0$ as $m \to 0$ and $\overline{\alpha} \to \infty$ as $m \to m_c$ with the expansion*

$$\overline{\alpha} \sim \frac{m_c}{m_c - m}. \tag{7.21}$$

The statements about the limits are taken from the paper [PZ95], who also showed partial monotonicity: $\overline{\beta} > 0$ if $m > m_s$, $\overline{\beta} < 0$ if $0 < m < m_s$. The analyticity and full monotonicity result are new and will be proved below.

We display in Figure 7.1 the results of the numerical computations done in [PZ95], and compare them with the ZKB exponents for $m > m_c$. We recall that in that case

[3] To be noticed that the space decay rate is twice as much.

$\alpha = 1/(m - m_c)$, a formula that is derived from invariance and mass conservation; since the last one fails here, $\overline{\alpha}$ is lower than the analogous formula, $\overline{\alpha} < 1/(m_c - m)$.

Let us record some consequences for the study of decay rates.

Proposition 7.3 *For all $m \in (0, m_c)$ the special self-similar solutions belong to $M^{p_*}(\mathbf{R}^n)$ and decay (in this norm) with the standard extinction rate*

$$\|\overline{U}(t)\|_{M^{p_*}} = c(T - t)^{-1/(1-m)}. \tag{7.22}$$

Moreover, we have $u(t) \in L^1(\mathbf{R}^n) \cap L^\infty(\mathbf{R}^n)$ with the estimate

$$\|\overline{U}(t)\|_{L^p} = c(T - t)^{\sigma_p}, \quad \sigma_p = \bar{\alpha} - (n\bar{\beta}/p), \tag{7.23}$$

for every $1 \le p \le \infty$.

The latter exponent must be positive for $p = 1$ since the L^1 norm of the solutions decreases in time, cf. Theorem 5.6. Here we recall the a priori restriction on the value of $\overline{\alpha}$: $0 < \overline{\alpha} < (n-2)/(n(m_c - m))$ (see (7.24) below), that implies a minimal rate for the decay of different norms. Curiously, this simplifies when translated in terms of L^1 decay, since it just means that $\sigma_1 > 1$. Thus, according to formula (7.23), it follows that for $0 < m < m_s$, the L^1-norm exponent $\sigma_1 \to 1$ as $m \to m_c$ and $\sigma_1 \to n/2$ as $m \to 0$.

7.2.1 Existence reviewed. Analyticity

We first review the fundamentals of the phase-plane analysis in line with similar analysis in previous chapters, and outline the proof of [PZ95]. We then present a proof that shows the analytic dependence of profiles and exponents on the parameters m and n.

(I) The self-similar profiles can be obtained by a phase plane analysis similar to the one stated in Sections 3.2, 7.1.1 and applied in Theorems 3.2 and 5.11, but more delicate. It boils down to finding a trajectory of the planar dynamical system connecting two specific singular points, called O and A in Subsection 3.2.2. This is referred to as *finding a correct connection.* Let us examine the situation in view of later results of this chapter. We are trying to construct radial Type-II self-similar solutions of the FDE equation for $0 < m < m_c$. We recall that when such solutions are globally defined in space and have no singularities, they are represented in the standard (X, Z) plane as orbits that start from $(0, 0)$ and continue into the upper half-plane (the second quadrant, to be precise), ending either at the singular point A, where $X = -(n-2)/m$ and $Z = 0$, or at the point B where $X = -2/(1-m)$ and $Z = Z_B = -Y_B > 0$. There is a third possibility for such orbits, namely that $X \to -\infty$ with $Z > 0$. Closer inspection shows that the profiles $f(\xi)$ of orbits in the third class cut the axis at a finite distance $\xi_0 < \infty$, and are not therefore acceptable as solutions of the Cauchy problem with non-negative data (they are changing sign solutions).[4]

[4] The existence of three possible types of orbits (slow, fast, and changing sign) is a very typical feature of nonlinear diffusion, cf. e.g., [FV03].

A further step in the analysis leads to the observation that A is a hyperbolic point, hence there is only one direction entering it. Repeating the local analysis of Proposition 5.12 in this situation, we see that there corresponds to it a unique orbit entering A from the second quadrant of the (X, Z) plane. Only for $\gamma > (n-2)/m$ or $\gamma < 0$ the slope is positive and the correct connection is in principle possible. Indeed, a qualitative analysis of possible dynamics easily shows that a correct connection is impossible when $0 < \gamma < (n-2)/m$, which means that necessarily

$$0 < \overline{\alpha} < \frac{n-2}{n(m_c - m)}. \tag{7.24}$$

Also O is a hyperbolic point and there is a unique orbit exiting from it; when $\alpha > 0$, it enters the second quadrant of the (X, Z) plane, thus allowing for a correct connection in principle.

The coincidence of those two branches is equivalent to the existence of the correct solutions described in the theorem. Showing that such a coincidence takes place for one and only value of $\overline{\gamma}$ and corresponding $\overline{\alpha}$ and $\overline{\beta}$ (the anomalous exponents) is the difficult part of the proof.

For every m the correct solution behaves like $O(|x|^{-(n-2)/m})$, which is called *fast spatial decay*. The uniqueness of the correct trajectory is proved by an analysis of the perturbation of the branches emanating from A and O when we change m and γ so that the connection is kept. This is done in detail in [PZ95].

(II) Here, we prefer to present a proof that provides us with the analytic dependence of α and f on m and n. We will proceed in several steps:

(i) We start from system (7.14) and change the horizontal variable, $X = -x/m$, so that the end-points of the connection do not depend on the parameters m and γ when expressed in terms of (x, Z). We also use the parameter $\mu = 1/\gamma$. Then, the system becomes

$$\begin{cases} \dot{x} = x^2 - (n-2)x + \frac{1}{1-m-2\mu}(m - \mu x)Z := F(x, Z, m, \mu, n), \\ \dot{Z} = \frac{1}{m}(2m - (1-m)x)Z := G(x, Z, m, \mu, n). \end{cases} \tag{7.25}$$

We are interested in finding solutions of this problem as curves $Z = Z(x)$ defined for $I = \{0 \le x \le x_1 = n-2\}$, with $Z = 0$ at the end-points and $Z > 0$ inside I.

(ii) We want to start from the explicit solution (7.19) for $m = m_s$ that corresponds to the values[5] $\mu = 0$ and

$$x(r) = \frac{(n-2)\, e^{2r}}{4n + e^{2r}}, \quad Z(r) = \frac{16n^2 e^{2r}}{(4n + e^{2r})^2},$$

hence

$$Z = ax(n-2-x) \quad \text{with } a = 4n/(n-2)^2. \tag{7.26}$$

[5] In that example, $\beta = 0$, hence $\gamma = \infty$, so that $\mu = 0$.

It is a parabola[6] passing through $P_0 = (0, 0)$ and $P_1 = (n-2, 0)$ with maximum at $x = (n-2)/2$. In the sequel $n > 2$ will be kept fixed, so that dependence on it will be omitted. Note that in the original variables the curve reads

$$Z = am\, X\, (n - 2 + mX).$$

(iii) In order to get a continuum of solutions, we will vary the parameters m and μ and consider the orbits starting from P_0 and P_1. We know that both end-points are hyperbolic points and the slopes of the orbits starting from them are

$$c_0 = \frac{n}{m\alpha} = \frac{n(1-m-2\mu)}{m} > 0, \quad c_1 = -\frac{(n-2-2m)(1-m-2\mu)}{m(m-(n-2)\mu)} < 0.$$

We note that $\partial c_0/\partial\mu < 0$ and $\partial c_1/\partial\mu < 0$. If we can continue those curves, and for some value of m and μ they meet at an intermediate point $x_* \in (0, n-2)$, then the correct connection will be obtained.

We will find such correct pairs m and μ as a function $\mu = \mu(m)$ by using the implicit function theorem. This is done as follows.

(iv) We write the system as

$$\frac{dZ}{dx} = H(x, Z, m, \mu) = \frac{G(x, Z, m, \mu)}{F(x, Z, m, \mu)}.$$

We immediately check that

$$\frac{\partial H}{\partial \mu} = -\frac{(2m-(1-m)x)^2 Z^2}{mA^2(1-m-2\mu)^2} \le 0,$$

where we have put $A = x^2 - (n-2)x + \frac{1}{1-m-2\mu}(m - \mu x)Z$ for brevity. Actually, $\partial H/\partial\mu = 0$ only at $x = x_B$.

(v) We now start the orbits at $m \sim m_s$, $\mu \sim 0$ and for small variations we have branches in the two parameters, $Z_l(x; m, \mu)$ starting from $x = 0$ and $Z_r(x; m, \mu)$ starting from $x = n-2$ (subscript l and r refer to left and right branch). They are prolonged at least until x_B. At that point, the following matching error happens:

$$E(m, \mu) = Z_r(x_B, m, \mu) - Z_l(x_B, m, \mu).$$

Since the condition for a correct connection is just $E(m, \mu) = 0$, and E is an analytical function of both arguments, if we can prove that $\partial E/\partial\mu \neq 0$, then the implicit function theorem will provide the branch $m = m(\mu)$ that we are looking for. Moreover, the solution will be locally unique. Now, we have

$$\frac{\partial E}{\partial \mu} = \frac{\partial Z_r}{\partial \mu}(x_B, m, \mu) - \frac{\partial Z_l}{\partial \mu}(x_B, m, \mu).$$

[6]The ZKB solution is also a parabola when expressed in the pressure variable $v = u^{m-1}$ as a function of x/t^β. However, it is a linear function in phase plane variables, cf. Subsection 3.2.4 (iii).

(vi) We will write the differential equation satisfied by $w_r = \partial Z_r/\partial\mu$ and $w_l = \partial Z_l/\partial\mu$ and show that $w_r > 0$ while $w_l < 0$ at $x = x_B$. That would complete the proof. The equation for w_i is

$$\frac{dw}{dt} = H_Z(x, Z)w + H_\mu(x, Z),$$

and the boundary conditions are

$$w_r(n-2) = 0, \quad w_l(0) = 0.$$

Due to the known slope behaviour we know that $w_r > 0$ for $x \sim n-2$ while $w_l < 0$ for $x \sim 0$. The lack of singularities in the rest of the trajectory allows us to conclude from the maximum principle.

(vii) We now need an argument to show that the branch can be continued for all $0 < m < m_c$. This relies on checking the geometric properties of the solutions that are being obtained. This is done in detail in [PZ95], where the asymptotics as $m \to 0$ and $m \to m_c$ are worked out. □

7.2.2 *The monotonicity result. Renormalized system*

We complete the information about the family of KPZ solutions by showing the monotonicity result of Theorem 7.2, that is evident from the numerical computation. The main idea is to perform a phase-plane analysis in *renormalized variables.*

Proof (i) We recall that the special solutions have trajectories connecting the critical point $O = (0, 0)$ in the (X, Z) phase plane to the critical point $A = (-(n-2)/m, 0)$; we also learn from the study of the dynamics of the flow in the plane that function $X(r)$ is decreasing for $-\infty < r < \infty$, and Z as a function of X is increasing for $-(n-2)/m < X < -2/(1-m)$ and decreasing for $-2/(1-m) < X < 0$. It will be convenient here to introduce new variables (sorry) to renormalize the situation so that the proof works. (Forgetting notations of the last proof) here we put

$$x = -mX/(n-2), \quad w = \alpha Z/nX, \tag{7.27}$$

so that the trajectories of the special solutions go from the point $P_0 = (0, -1)$ to the point $P_1 = (1, 0)$ in the new (x, w) phase plane. It will also be convenient to introduce the parameter $\mu = n/\gamma$ which will vary in the range $-\infty < \mu < 1$. Note that

$$\alpha = n/(n(1-m) - 2\mu), \quad \beta = \mu\,\alpha$$

are increasing functions of μ for fixed m and n. Condition (7.24) implies that we are only interested in μ's such that $\mu < nm/(n-2)$.

The flow is now controlled by the analytic system

$$\begin{cases} \dot{x} = -x\left((n-2)(1-x) + nw - \frac{(n-2)\mu}{m} xw\right) = F(x, w), \\ \dot{w} = w\left(n + nw - \frac{(n-2)}{m}(1+\mu w)x\right) = G(x, w), \end{cases} \tag{7.28}$$

where $\dot{x}$ means dx/dr. This system has critical points $O = (0,0)$, P_0, P_1, and $P_B = (x_B, w_B)$ with

$$x_B = \frac{2m}{(n-2)(1-m)}, \qquad w_B = -\frac{n(1-m)-2}{n(1-m)-2\mu},$$

that corresponds to the point $B = (X_B, Y_B)$ of the original system of Section 3.2.2 (note that x_B travels from 0 to 1 as m grows from 0 to m_c, while the evolution of w_B in $(-1, 0)$ depends on the relation between μ and m).

This system admits a remarkable explicit solution for the values $m = m_s = m_c/2$ and $\mu = 0$ $(\gamma = \infty)$,

$$w = -1 + x.$$

This is the explicit case of KZP solution to which we have referred before. It will appear again in Section 7.5.

(ii) We also need the behaviour of the system flow near P_0 and P_1. We already know that the latter is a saddle point, since it corresponds to point A in the original system that has been analysed in Proposition 5.13; using the variable $x' = x - 1$, the linearization is

$$\dot{x} = (n-2)x' - \left(n - \frac{(n-2)\mu}{m}\right) w, \qquad \dot{w} = -w\,\frac{n(1-m)-2}{m}.$$

It has eigenvalues $\lambda_1 = n - 2$ with eigenvector $(1, 0)$ and $\lambda_2 = -(n(1-m)-2)/m$ with eigenvector $(1, c)$ where

$$c(m, n, \mu) = \frac{n-2-2m}{nm-(n-2)\mu}.$$

Note some features: we have $c > 0$, $\partial c/\partial\mu > 0$ for fixed m, and also $\partial c/\partial m < 0$ for fixed $\mu < n/2$ (as is the case in our application, where $\mu < nm/(n-2) < 1$). Also $\partial c/\partial n > 0$. Along the curve of anomalous exponents $\mu = \overline{\mu}(m)$ we have $c \to 0$ as $m \to 0$ and $c \to \infty$ as $m \to m_c$.

According to the general theory of autonomous dynamical systems, the behaviour of the nonlinear system near a hyperbolic point is given in first approximation by the linearization, see [GH83] or [Pk91]. In particular, there exists a unique orbit that enters P_1 and it follows the direction of the eigen-vector. The orbit depends in an analytic way on the parameters.

The linearized analysis of P_0 is as follows. Putting $w = -1 + w'$ we have

$$\dot{x} = 2x, \qquad \dot{w}' = \frac{(n-2)(1-\mu)}{m}x - nw'.$$

It is again a saddle point with eigenvalues $\lambda_1 = -n$ and $\lambda_2 = -2$ and respective eigenvectors $(0, 1)$ and $(1, d)$ with

$$d(m, n, \mu) = \frac{n-2}{m(n+2)}(1-\mu).$$

The orbit that gets out of P_0 travels along this last direction. We have $d > 0$, $\partial d/\partial\mu < 0$, $\partial d/\partial m < 0$, $\partial d/\partial n > 0$, and $d(m_c, n, 1) = 0$.

(iii) We are interested in localizing the regions of the plane that allow for connections from P_0 to P_1. In this respect, it is useful to consider the locus of vertical slopes, where $\dot{x} = 0$, that is composed of the axis $x = 0$ and the line L_1:

$$w = -\frac{(n-2)(1-x)}{n - \frac{(n-2)\mu}{m} x},$$

as well as the locus of horizontal slopes where $\dot{w} = 0$, composed of the axis $w = 0$ as well as the line L_2:

$$x = \frac{nm(1+w)}{(n-2)(1+\mu w)}.$$

We are interested in the rectangle $\mathcal{R}_1 = \{(x, w) \in (0, 1) \times (-1, 0)\}$, where the KPZ solutions lie. In that region both lines cross exactly at the point P_B. A topological analysis of the flow in this rectangle shows that the unique trajectory emanating from P_0 has three options:

(a) Enter P_1, and this would be a KPZ connection.

(b) Cross the line L_1 and then L_2 and return towards the origin for a while around the point P_B. There will be no connection in that case, but a focusing solution (either converging to P_B or having a limit cycle; this depends on m and μ).

(c) the orbit exits $\mathcal{R}_1$ through the lateral boundary $x = 1$ and never returns. Since orbits cannot cross, solutions of the three types are ordered as long as they are monotone, and this happens below the joint-line L_{12}.

The choice of one of these possibilities depends on m and μ. A similar analysis applies to the orbit entering P_1, and three classes appear. We skip the details.

(iv) We now address the question of monotonicity of $\overline{\alpha}$ with $m \in (0, m_c)$. If the property is not true, in view of the limit values of $\overline{\alpha}$ near the end-points already established, we must be able to find values $m_1 < m_2$ with the same value of $\overline{\alpha} = \alpha_0$.

We now have the following claims: (i) the solutions $w_i(x)$, $i = 1, 2, 3$, must be ordered near the end-points: we have

$$w_1(x) < w_2(x) < 0$$

near $x = 1$ with $x < 1$, and the order is reversed for small increments near $x = 0$, $x > 0$,

$$w_1(x) > w_2(x) > -1.$$

This follows from the analysis of the orbits emanating from the end-points P_0 and P_1. It follows that the curves $w = w_1(x)$ and $w = w_2(x)$ must intersect somewhere in the middle.

However, we can show that no intersection is possible. At the first intersection, say (x_0, w_0), we must have $dw_1/dx \le dw_2/dx$, i.e., G/F increasing with m. But if we inspect the variation of the vector field (F, G) with m, we discover that

$$\frac{\partial F}{\partial m} = -\frac{n-2}{m^2} x^2 w > 0, \qquad \frac{\partial G}{\partial m} = \frac{(n-2)}{m^2} x(1+\mu w) w < 0.$$

so that

$$\frac{\partial(G/F)}{\partial m} = \frac{1}{F^2}(G_m F - F_m G) < 0.$$

This ends the proof. It is like magic.[7] □

7.2.3 Phenomenon of relative concentration for $m < m_s$

An interesting aspect of Theorem 7.2 is the interpretation in terms of how the solutions spread and what is the role of exponent m_s in the classification of fast diffusions.

When we look at the range $m_s < m < m_c$ we see that the special self-similar solutions have a positive exponent β which means the mass distribution they represent has a tendency to spread in time. We have to be precise in what we mean since we also have loss of mass (and eventual extinction). A way of measuring such spread is as follows: for every $\lambda \in (0, 1)$ the line where $u(x, t)$ takes the intermediate value $\lambda u(0, t)$ evolves with time as

$$|x| = x_0(\lambda)(T - t)^{-\beta}, \tag{7.29}$$

and this curve goes to infinity as $t \to T$ since $\beta > 0$. This phenomenon goes in line with what is known to happen for $m \geq m_c$, where solutions also spread, although the rate is different.

A curious novelty appears when $0 < m < m_s$, since then $\beta < 0$, and the λ-lines of the self-similar distribution converge to the origin, $x = 0$, so that after normalization of the mass to one, we can say that the (relative) mass distribution tends to concentrate at $x = 0$ in the form of a (small) Dirac mass. Indeed, a very strange thing happens since the mass disappears anyway, and this may only happen through $x = \infty$ as we have explained in Section 5.5.1. We give to this phenomenon the illuminating name of *relative concentration* at extinction.

An alternative (and more elegant) way of calculating the expansion (or relative concentration) is to calculate the mean deviation of the mass distribution $u(\cdot, t)$, which is given by the formula

$$\overline{x}(t)^2 = \frac{\int x^2 u(x, t)\, dx}{\int u(x, t)\, dx}. \tag{7.30}$$

It is quite easy to see that the mean deviation of the self-similar special solutions is finite precisely for $m < m_s$ (use the asymptotic behaviour $u \sim c(t)|x|^{(2-n)/m}$ to prove it), and the formula for the self-similar mean deviation is the same, $\overline{x}(t) = \overline{x}_0(T - t)^{-\beta}$, so that it goes to zero as $t \to T$.

Situated in between these opposite cases, the case $m = m_s$ exhibits separation of variables and the λ-lines are constant in time. Extinction is accompanied by a relative stabilization in the standard space variable.

---

[7]The reader could complain that the pace of proof is maybe too brisk, but similar ideas are explained in full detail in [AV95] applied to the focusing problem for the PME.

7.3 Admissible extinction rates

Let us deal a bit more with the phase-plane analysis in order to recover other radial Type-II self-similar solutions that also play a role in a full description of extinction. We have explained after the statement of Theorem 7.1 the existence of three possible types of orbits in the phase-plane emanating from $(0, 0)$. The first class consisted of the 'correct' connections from O to A, giving rise to self-similar solutions with fast decay. As for the second class, i.e., the orbits going to B, the solutions have a behaviour as $x \to \infty$ like $O(|x|^{-2/(1-m)})$. This is called slow spatial decay. These types of connections from 0 to B provide us with examples of self-similar solutions of the FDE in the range $0 < m < m_c$ that have a finite extinction time and exhibit a quite definite extinction rate and profile. The orbits in the third class reach the level $f = 0$ at a finite distance and do not provide solutions of the FDE in the whole space. Though these orbits are in principle useless for the theory of non-negative solutions of the Cauchy problem, they are subsolutions and we will find a use for them below.

We want to classify the situation of the three classes in terms of α. This is the result.

Theorem 7.4 *For every $\alpha \in (\overline{\alpha}, \infty)$ there is a self-similar solution of the FDE of the form*

$$u(x,t) = (T-t)^{\alpha} f(x(T-t)^{\beta}) \tag{7.31}$$

with $\beta = (\alpha(1-m)-1)/2 > \overline{\beta}$, and $f = f_\alpha$ has a slow spatial decay at $x = \infty$. For $\alpha = \overline{\alpha}$ there exists a unique solution with fast decay, and for $0 < \alpha < \overline{\alpha}$ no global profile f_α exists, but there is a radially symmetric profile that cuts the axis at a finite distance. In all cases, f_α is radial and decreasing.

Proof We only need to shoot from the origin, take into account the local behaviour of the orbits in their dependence of the parameters, and recall that the entrance into A happens only for $\gamma = \overline{\gamma}$ when m and n are kept fixed. Looking for instance at the local analysis near O done in Proposition 5.12 we see that the slope of the outgoing orbit is

$$c_0 = \frac{n}{\alpha},$$

which is positive for $\alpha > 0$. As for point A, Proposition 5.13 implies that the slope of the entering orbit is

$$c_1 = -\frac{n-2-2m}{\alpha(m-\frac{n-2}{\gamma})}.$$

There are two possibilities: the slope is negative if $\gamma > (n-2)/m$ or $\gamma < 0$, and it is positive otherwise, i.e., when $0 < \gamma < (n-2)/m$. We exclude the range $0 < \gamma < 2/(1-m)$ since α is negative in that case. For $\gamma = 2/(1-m)$ we cannot use the formulas since α is not defined. On the other hand, for $\gamma = (n-2)/m$ we have a vertical slope.

Let us now see what happens with the orbits emanating from O in the two cases. First of all, in the range $\gamma \in (2/(1-m), (n-2)/m)$, an easy qualitative analysis shows that the orbit emanating from 0 cannot enter A (it would have to enter from a

SW direction which means crossing before the line $X = -(n-2)/m$, and this is not possible). It therefore belongs to the second class, which revolves around B.

The same is true for γ a bit larger than $2/(1-m)$ by continuity. If we continue increasing γ, i.e., decreasing α, the orbits will belong to the same class until one of them becomes a correct orbit and enters A. This follows from continuity arguments once the critical points and regions of growth and decrease of the orbits are laid out. We leave the task to the reader. Summing up, when $\alpha > \overline{\alpha}$ the connection revolves around point B and the corresponding solution of the FDE has slow decay.

We still have to prove that for $\alpha < \overline{\alpha}$ the orbit from O goes to $X = -\infty$. This needs an analysis of the evolution of the branches coming out of O and A for $\alpha < \overline{\alpha}$. The idea is to go to the extreme situation where γ is negative and goes to zero so that $\alpha \to 0$ and $\beta \to -1/2$. Then we see that the branch coming from O has initial slope tending to ∞ and the branch coming into A has slope tending to a finite value. The analysis shows then that the latter moves on top of the former without meeting at a given point in the middle, say at $X = -2/(1-m)$. This implies that the orbit coming from O is an orbit of the third class. This will be the case when we vary α upwards until we come to a correct orbit (same argument as before). The correct orbit is unique, so that the argument ends. □

Let us summarize the result on different extinction rates that we have collected.

Corollary 7.5 *For every $\alpha \geq \overline{\alpha}(m,n)$ there exist self-similar solutions which decay near extinction with the rate*

$$\|u(t)\|_\infty = c\,(T-t)^{-\alpha}. \tag{7.32}$$

The region is represented in the diagram of Figure 7.2. All of these solutions belong to $M^{p_*}(\mathbf{R}^n)$ and decay in that norm with the standard extinction rate

$$\|u(t)\|_{M^{p_*}} = c(T-t)^{-1/(1-m)}. \tag{7.33}$$

Only in the case $\alpha = \overline{\alpha}(m,n)$ some of these solutions are integrable, $u(t) \in L^1(\mathbf{R}^n)$ for $0 < t < T$.

Figure 7.2 explains this information. The dashed line represents the singular solution exponent $\alpha = 1/(1-m)$ (only for $m < m_c$, for $m > m_c$ it does not correspond to an extinction time profile), while the lined region displays the allowed self-similar exponents with bounded profiles, limited from below by the curve of values of $\overline{\alpha}(m)$ which is asymptotic to $m = m_c$.

The following bound from below completes the information on admissible rates for radial solutions.

Theorem 7.6 *Let u be a bounded radial solution of the FDE with $0 < m < m_c$ which is non-increasing in the radial variable and has extinction time $T < \infty$. Then, for every $\alpha < \overline{\alpha}$ there is a constant $c(u_0, \alpha) > 0$ such that*

$$\|u(x,t)\|_\infty \leq c\,(T-t)^{\alpha}. \tag{7.34}$$

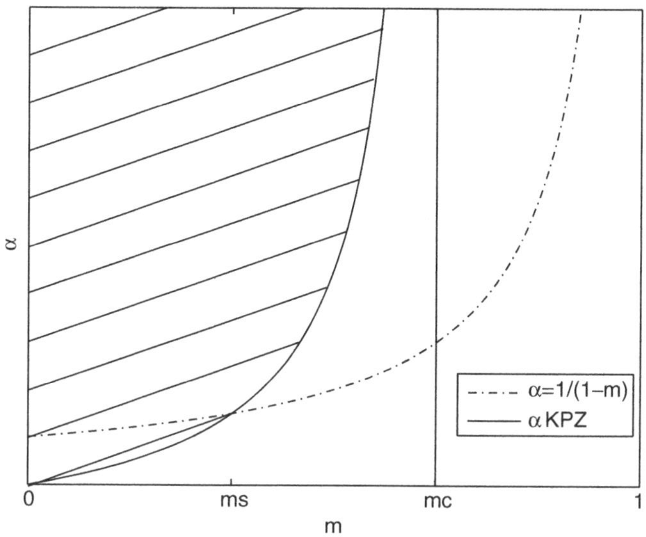

Figure 7.2 The allowed extinction exponents.

Proof Let T be the extinction time of the solution. The idea is to compare our solution with the self-similar solution $\tilde{u}$ with same T and parameter α that as discussed above has a profile that cuts the axis. By scaling we make it cut the axis very close to the origin, say at $\eta = |x|(T - t)^{\beta}$, and to be very high.

We now use the standard intersection number count (cf. e.g. [St69, Ma82, Ga04, V06]). We take as space-time domain for our argument (using radial variables in space) $Q = \{(r, t) : 0 < t < T, 0 < r < R(r)$ with $R(t) = \eta(T - t)^{-\beta}\}$. According to that theory, if we count the number of sign changes of the difference $z_t(r) = u(r, t) - \tilde{u}(r, t)$ at fixed times, that number (lap number) does not increase with time unless new intersections of the curves come into the domain from the boundary. Now, in the conditions of the previous construction that number is initially one. It is clear that no new sign changes come through the lateral boundaries (since $u > 0 = \tilde{u}$ at $r = R(t)$, and we have zero Neumann conditions at $r = 0$).

We conclude from the theory that the intersection number must be either 0 or 1 for $0 < t < T$. If for some $t < T$ the intersection is lost, it means that $u(t) \geq \tilde{u}(t)$ in Q, and we conclude by the strong maximum principle that the extinction time of u os larger than T, a contradiction. Therefore, there is an intersection all the time. By the symmetry and monotonicity assumption, the set where the difference has a plus sign must include $x = 0$, hence

$$\|u(t)\|_{\infty} \leq \tilde{u}(0, t) = C\,(T - t)^{\alpha}.$$

This ends the proof. □

We have proved that for such solutions

$$\liminf_{t \to T} \frac{\log \|u(t)\|_\infty}{\log(T-t)} \geq \overline{\alpha}. \tag{7.35}$$

After these results, we can propose the name of *minimal extinction solutions* for the solutions $\overline{U}$ with anomalous exponents of Theorem 7.1. More precisely, they have minimal similarity exponents and also minimal shape as $x \to \infty$.

7.4 Radial asymptotic convergence result

We now address the question of asymptotic extinction behaviour. We want to show that the family of special solutions $\overline{U}$ with anomalous exponents are actually the asymptotic extinction profiles to which a large class of solutions evolves near extinction. That would place those solutions in the same position the ZKB solutions are in the upper range $m > m_c$.

The result is known under the assumption of radially symmetric and integrable data, and without that restriction when $m = (n-2)/(n+2)$. The radial result was first shown in a formal way by King [Ki93], who obtained a fairly detailed description of the vanishing profiles for radially symmetric solutions using a formal matching asymptotic analysis. The result was rigorously proved by Galaktionov and Peletier in [GP97] under a number of assumptions. We improve those assumptions using the results of previous chapters as follows.

Theorem 7.7 *Let $u(|x|,t)$ be a radial solution of the Cauchy problem for the FDE in dimensions $n \geq 3$ with $0 < m < m_c$. Assume that the initial data satisfy $u_0 \in L^{p_*}(\mathbf{R}^n)$ and*

$$u_0(r) = O(r^{-q}) \quad as\ r \to \infty \tag{7.36}$$

for some $q > 2/(1-m)$. Let $T = T(u_0) > 0$ be the finite extinction time. Then, there exists a constant $a_0 > 0$ depending on u_0 such that

$$(T-t)^{\bar{\alpha}}\, u(\eta(T-t)^{\bar{\beta}}, t) \to F(\eta; a_0) \tag{7.37}$$

as $t \to T$ uniformly with respect to $\eta \geq 0$.

Theorem A of paper [GP97] proves this result under the additional assumptions $u_0 \in L^1(\mathbf{R}^n) \cap L^p(\mathbf{R}^n)$ with $p > 2(1-m)/n$ and

$$u_0(r) = O(r^{-(n-2)/m}) \quad \text{as } r \to \infty. \tag{7.38}$$

After introducing the rescaled solutions of the form (7.9), which satisfy equation (7.10), the proof is based on a geometric Lyapunov-type argument and comparison arguments based on the intersection properties of the solution graphs. We refer to this work for a complete understanding of the long argument.

Here are our reduction steps: after the results of Chapter 6, the assumption $u_0 \in L^{p_*}(\mathbf{R}^n)$ implies that $u(t) \in L^\infty(\mathbf{R}^n)$ for $t > 0$. Finally, the asymptotic condition

(7.36) implies (7.38) for all later times by comparison with the self-similar solution of Theorem 5.14 (the backward-effect solution).

Let us now discuss the optimality of the assumptions. The exponent in assumption $u_0 \in L^{p_*}$ cannot be lowered since we know for all $p < p_*$ there are solutions with a standing singularity, namely the self-similar solutions for the backward effect. Nor can be replaced L^{p_*} by M^{p_*} for the same reason. But note that Chapter 6 allows us to use initial data $u_0 \in M^{p_*}(\mathbf{R}^n)$ with a Marcinkiewicz constant $\mathrm{N}_p(u_0)$ small with respect to $T^{1/(1-m)}$, and then the solutions would become bounded before the blow-up, thus entering into the assumptions of the already proved result.

Note finally that the exponent in (7.36) cannot be relaxed because the assumption $u_0(r) = O(r^{-2/(1-m)})$ does not fall inside $L^{p_*}(\mathbf{R}^n)$. On the other hand, we conjecture that solutions with such a behaviour do not satisfy the asymptotics described in the theorem.

A similar convergence result is conjectured to be true without the assumption of radial symmetry. Such an extension of the previous theorem is only known in the special case of exponent $m_s = (n-2)/(n+2)$, where the equation has a special structure.

7.5 FDE with Sobolev exponent $m = (n-2)/(n+2)$

The exponent is special in the sense that when we separate variables and pass to the Emden–Fowler equation, then $p = 1/m$ equals the Sobolev exponent $(n+2)/(n-2)$, that is known to have a big influence on the qualitative solutions of that elliptic equation. We will call m_s the Sobolev exponent of the FDE.

7.5.1 *The Yamabe flow. Inversion and regularity*

The FDE is related for this precise exponent to the famous Yamabe flow of Riemannian geometry. This flow is used by geometers as a tool to deform Riemannian metrics into metrics of constant scalar curvature within a given conformal class. We review the main facts about the Yamabe problem for the reader's convenience in Appendix III.

Preparing the way for the analysis of extinction, we note that the anomalous similarity exponent α is now $(n+2)/4$ (and corresponds to separate variables, $\beta = 0$). The corresponding family of self-similar solutions is explicitly known and given by the Loewner–Nirenberg formula (7.19) that we can also write in the form (cf. [LN74])

$$F(x,\lambda) = k_n \left(\frac{\lambda}{\lambda^2 + |x|^2} \right)^{(n+2)/2} \tag{7.39}$$

with $k_n = (4n)^{(n+2)/4}$ and $\lambda > 0$ arbitrary. They represent conformal metrics with constant curvature in the interpretation associated to the Yamabe problem, as we explain in Appendix III. In fact, it is well known that such a metric corresponds to the stereographic projection from the space $\mathbf{R}^n$ into the sphere $\mathbb{S}^n$.

Let us also introduce a technical tool that we will need presently, the inversion. It is given by the formula

$$\widetilde{u}(x,t) = |x|^{-(n+2)} u(I(x),t), \quad I(x) = x/|x|^2. \tag{7.40}$$

It can be checked after some work that whenever u is a solution of the FDE with $m = (n-2)/(n+2)$, then $\widetilde{u}$ is again a solution of the same equation. The inversion can be applied to the profiles of the Emden–Fowler equation with $p = p_s$ with equal results. In particular, F_λ transforms in this way into $F_{1/\lambda}$.

Another quite important property of inversion is the conservation of the L^{p_*} norm of the function being inverted, where $p_* = n(1-m)/2 = 2n/(n+2)$. Indeed,

$$\int \widetilde{u}^{p_*}(x,t)\,dx = \int |y|^{(n+2)p_*} u(y,t)^{p_*} |y|^{-2n} dy = \int u(y,t)^{p_*} dy,$$

i.e., $\|u(t)\|_{p_*} = \|\widetilde{u}\|_{p_*}$.

The inversion formula gives us the possibility of interchanging the behaviour near $x = 0$ with the behaviour as $|x| \to \infty$. This fact leads to the following consequence of Theorem 6.7, an improvement in time of the behaviour at infinity.

Theorem 7.8 *Let* $n \geq 3$, $m = m_s$, $u_0 \in L^{p_*}(\mathbf{R}^n)$ *and let* u *be the solution of the FDE. Then for all* $t > 0$ *we have* $u(t) \in L^\infty(\mathbf{R}^n) \cap L^1(\mathbf{R}^n)$ *and moreover* $u(t) \leq C(t)|x|^{-n+2}$ *for some finite function* $C(t)$.

For the proof, observe that $u_0 \in L^{p_*}(\mathbf{R}^n)$ implies $\widetilde{u}_0 \in L^{p_*}(\mathbf{R}^n)$, hence $\widetilde{u}(t)$ is bounded for all $t > 0$ by Theorem 6.7. Using again the inversion, this implies that

$$u(x,t) = O(|x|^{-(n+2)}).$$

Remark Recalling that for $m = m_s$ we have $n+2 = (n-2)/m$, we see that this result improves the backward effect estimate of Chapter 5. We have been unable to establish such a result for $m \neq m_s$.

7.5.2 Asymptotic behaviour

Here is the result proved by Del Pino and Saez, where the condition of radial symmetry is eliminated with respect to the results of [GP97]. Let us define the weighted norm

$$\|f\|_* = \sup_x \,(1 + |x|^{n+2})|f(x)| \tag{7.41}$$

for functions $f \in L^1_{\mathrm{loc}}(\mathbf{R}^n)$. Boundedness of this norm means a fast decay condition similar to the one imposed by [GP97].

Theorem 7.9 [DPS01] *Let* $u(x,t) \geq 0$ *be a non-trivial solution of the Cauchy problem for the FDE in dimensions* $n \geq 3$ *with* $m = (n-2)/(n+2)$. *Assume that the initial function* u_0 *is continuous and* $\|u_0\|_*$ *finite. Let* T *be the vanishing time of the solution. Then, there exist* λ *and* $x_0 \in \mathbf{R}^n$ *such that*

$$(T-t)^{-(n+2)/4} u(x,t) = F(x - x_0, \lambda) + \theta(x,t) \tag{7.42}$$

and $\|\theta(t)\|_* \to 0$ *as* $t \to T$.

The proof is done by rescaling the solution as in (7.6) and then lifting the solution to the sphere $\mathbb{S}^n$ by means of the formula

$$v(y,\tau) = (1+|x|^2)^{(n+2)/2}(T-t)^{-(n+2)/4}\,u(x,t) = k_n\,\frac{u(x,t)}{U(x,t)} \qquad (7.43)$$

where U is the explicit solution (7.20), $y = \pi(x)$ is the stereographic projection of $x \in \mathbf{R}^n$ into $\mathbb{S}^n$ and $\tau = \log(T/(T-t)) \in (0,\infty)$. It is then shown that v satisfies the following semilinear parabolic equation on the sphere (cf. [DPS01, Ye94])

$$v_\tau = \Delta_S v^m - c_n v^m + \frac{1}{1-m}v, \qquad (7.44)$$

posed for $(y,s) \in \mathbb{S}^n \times (0,\infty)$; Δ_S denotes the Laplace–Beltrami operator in $\mathbb{S}^n$ and c_n is a positive constant.

It is now clear that what we are stating is a result of uniform convergence on the sphere of the function corresponding to F (i.e., θ becomes in this setting a uniform small relative error).

We combine the two last theorems into a result that considerably relaxes the assumptions on u_0 for the FDE with Sobolev exponent. Indeed, it shows a first approximation to the basin of attraction of F for the Yamabe flow.

Theorem 7.10 *Let $m = (n-2)(/n+2)$. Under the sole assumption that $u_0 \in L^{2n/(n+2)}(\mathbf{R}^n)$ the solution of the FDE is bounded for all $t > 0$, $\|u(t)\|_*$ is finite, and the vanishing behaviour of Theorem 7.9 holds. The exponent is optimal.*

Proof First of all, under this assumption we know by the results of Chapter 6 that $u(x,t)$ is bounded, hence continuous, for all positive t. So we may assume that u_0 is continuous from the outset. Next, we use Theorem 7.8 to fall into the assumptions of Theorem 7.9.

As for optimality, there are unbounded solutions in $L^p_{\rm loc}(\mathbf{R}^n)$ with $p < p_*$, as the ones constructed in the backward effect, cf. Theorem 5.11. □

7.6 The Dirichlet problem in a ball

As a complement to the previous discussion of the problem in the whole space, we consider in this section the evolution of the solutions of the FDE with exponents $0 < m < 1$ posed in a ball $\Omega = \{|x| < 1\} \subset \mathbf{R}^n$, $n \geq 3$, with boundary $\Gamma = \{|x| = 1\}$. We take boundary conditions

$$u(x,t) = 0 \quad \text{on } \Gamma \times (0,T), \qquad (7.45)$$

and initial conditions $u(x,0) = u_0(x)$ for $x \in \Omega$. We assume that u_0 is bounded, continuous and non-negative. Known theory guarantees that this Dirichlet problem admits a unique classical solution for a certain positive time. The solution is known to vanish in finite time $T > 0$, so that $u > 0$ in $\Omega \times (0,T)$ and $u(x,T) = 0$.

The asymptotic behaviour of such solutions depends on whether $0 < m < m_s$, $m = m_s$ or $m_s < m < 1$. The latter case is usually called subcritical in the literature, thinking in terms of the associated elliptic problem, but should be termed as supercritical in the notation followed in this text. Such case was solved by Berryman and Holland in a celebrated paper in 1980 [BH80]. They prove that solutions converge after rescaling to the self-similar solution in separated variables, which amounts to finding the profile of the Emden–Fowler equation in the subcritical range $1 < p < p_s$, defined in a bounded domain with zero boundary conditions.[8]

In the supercritical case (subcritical for us) $0 < m \le m_s$ such smooth separate-variable solutions do not exist. Galaktionov and King study the problem in [GK02] with positive symmetric initial data $u_0(r)$, $r = |x|$, and show that the asymptotic extinction behaviour is described by matched asymptotics. For the study of the inner region, near $x = 0$, they use the fact that the self-similar solution with anomalous exponents (which they call minimal self-similar solution of the second kind) is asymptotically stable as $t \to T$ in the symmetric geometry as we have discussed in Section 7.4. Recall that the solution is radially symmetric and has the form

$$U(x,t) = (T-t)^{\alpha} F(|x|(T-t)^{\beta}),$$

and the anomalous exponent β is negative. They prove that in an inner region of size $|x| = O((T-t)^{|\beta|})$, (which shrinks to a point) the solution u is described to leading-order by U as $t \to T-$,

$$u(x,t) \sim U(x,t).$$

In any case, this region marks the L^{∞} norm of u as $t \to T$. On the other hand, the outer region is $|x| = O(1)$ with

$$u(x,t) \sim (T-t)^{\mu}\Phi(x),$$

where Φ is the Green function satisfying $\Delta\Phi(x) = -C\delta(x)$ in Ω, $\Phi = 0$ on $\partial\Omega$, and $\mu = \alpha - \beta(n-2)/m$.

The same authors show that in the critical Sobolev case $m = m_s$ the asymptotic behaviour as $t \to T-$ is given by matched asymptotics: near the origin $x = 0$ is essentially non-self-similar and is constructed by matching the expansions in the inner and outer domains. This gives the extinction rate as $t \to T-$:

$$\|u(\cdot,t)\|_{\infty} = \gamma_0 (T-t)^{(n+2)/4} |\ln(T-t)|^{(n+2)/2(n-2)}(1+o(1)), \tag{7.46}$$

where $\gamma_0 = \gamma_0(n) > 0$ is a constant. Note that $(n+2)/4 = 1/(1-m_s)$ is the 'natural exponent' found before; it comes affected by a logarithmic correction.

7.7 Comments, extensions, and historical notes

Intermediate asymptotics We have found that the absence of the ZKB solutions has been compensated by the existence of a new family of self-similar solutions that allow

[8] Such a profile need not be unique if Ω is not a ball.

us to state the relevant asymptotic convergence results for solutions with L^1 data. We realize the crucial role played in the long-time dynamics of these nonlinear heat equations by the existence of certain types of particular solutions.

Our analysis has been focused on integrable data. However, our previous results show that all solutions with exponents (m, p) below the critical line fall at positive times into the class of solutions with L^1 data. We have even seen that some solutions with strongly singular data do the same.

Section 7.1 Self-similarity of Type II can be considered for all $m \in \mathbf{R}$, but plays a significant role in the study of extinction for the FDE with subcritical exponent, as we have just seen. The reader is warned that he self-similar form (7.1) is found in the literature under a number of variants, like $U(x, t) = t^{\alpha} f(x\,(T-t)^{-\beta})$ cf. [PZ95]. The signs of α and β chosen here is intended to make easier the comparative analysis of both types of solutions via phase-plane analysis. The idea of the plane reflection to represent both types in one plane is taken from [LOT82].

Section 7.2 The main result, Theorem 7.1 is taken from [PZ95] (with somewhat different notation). I thank M. Peletier for the numerical computations of Figure 7.1(b).

The best-known anomalous exponent problem in nonlinear diffusion is maybe the focusing problem for the PME studied by Aronson and Graveleau in [AG93].

The analyticity is new but uses the ideas of [AV95], where the case of focusing solutions of the PME is treated.

The monotonicity result of Subsection 7.2.2 is new. It would be nice to have a PDE proof of the monotonicity of the minimal exponents in n and m. We guess this would be possible if some new idea is introduced. Another open problem is to prove the monotone dependence of α on n. In the case of the focusing problem for the PME, the limit behaviour was proved in [AGV98]: $\beta_*(m, N) \to 1/2$ if $m \to \infty$, while it tends to 1 as $m \to 1$, always for $N \geq 2$. The monotonicity of β_* as a function of m in that problem was proved in [ABH03].

Section 7.3 The results about different extinction rates seem to be new. The existence of bounded separable solutions (case $\alpha = 1/(1-m)$, $\beta = 0$) was proved by Berestycki, Lions and Peletier [BLP81].

A whole analysis of the question of actual decay rates in different norms when $m < m_c$ is still in progress.

We conjecture that Theorem 7.6 is true under weaker assumptions on the data. Can the reader eliminate at least the assumption of monotonicity?

Section 7.4 The main result of Theorem 7.7 is taken from [GP97], and the application to general data is new, a consequence of our results in Chapter 5.

Open problem The extension of these results to non-radial data is completely open when $m \neq m_s$. In fact, it is not clear that the radial profiles with anomalous exponents will be attractive in all cases. One guess is that they are attractive for $m \in [m_s, m_c)$ but not always for $0 < m < m_s$. We find that solving this question is an important open problem. Note that such non-radial instabilities are typical of continuum mechanics; more to the point, they have been found in the closely related study of anomalous exponents for the focusing problem in the PME, cf. [AA95, AA01].

Section 7.5 The connections with Geometry are a classical subject. The study of the elliptic Emden–Fowler problem in the Sobolev case was initiated by Brezis and Nirenberg in [BN83] and followed by many authors.

Theorem 7.9 is taken from [DPS01]; we improve the class of initial data to an almost optimal class using Chapters 5 and 6 and the inversion transformation.

Section 7.6 For the classical (local-in-time) existence theory see, e.g., the references in [Kw98]. The property of extinction in finite time was first proved by Sabinina [Sb62, Sb65] for a class of one-dimensional parabolic equations of fast diffusion type in bounded intervals.

The proof of the exact matched asymptotics in [GK02] is a very delicate work. The existence of logarithmic corrections is a very peculiar feature of dynamical systems in critical situations. An early case example in the area of nonlinear diffusion was worked out by Galaktionov and the author in [GV91].

The extension of the results for non-radial solutions is an interesting open problem. For the case $m > m_s$ the problem is addressed in Bonforte and Vázquez [BV06], where convergence to steady profiles in relative error norm is established.

Duality The existence of the inversion mapping for the FDE with Sobolev exponent implies that the forward and the backward effect of Theorems 5.1 and 5.11 are somewhat equivalent. There seems to be a certain duality between smoothing into L^∞ and improved decay as $x \to \infty$. We have used the smoothing effect of Chapter 6 to get a more precise result about improved decay at infinity, i.e., the conclusion that $u(t)$ decays like $O(|x|^{-(n+2)})$.

Open problem Investigate the existence of some form of the above-mentioned duality for $m \neq m_s$, $0 < m < m_c$.

Admissible rates In order to fix ideas, let us take as standard decay the one with exponent $\alpha = 1/(1-m)$ as given in local average by the explicit solution (5.2); we see that for $m = m_s$ we have standard and slower than standard decays, for $m > m_s$ only slower than standard, for $m < m_s$ the whole range of all rates faster than standard, plus the standard and some slower.

Debate Note that the concepts of fast or slow decay near an extinction point are somewhat confusing. A higher power of $T - t$ in the decay rate (the α in the self-similar solutions) implies in principle a less abrupt form of approach to zero which can be seen as slower rate. On the other hand, once T is fixed a higher power of $T - t$ means a smaller function, which could be seen as faster rate. Since considering T fixed is a less likely assumption, we favour the first alternative.

Exercises

Exercise 7.1 This is an exercise about the explicit solution of formula (7.19) for $m = m_s$. Show that in that case

$$x(r) = \frac{\eta^2}{4n + \eta^2}, \qquad w(r) = -\frac{4n}{4n + \eta^2}$$

with $\eta = e^r$, so that system (7.28) has the explicit solution $w = -1 + x$ for $m = m_s$ and $\mu = 0$.

Exercise 7.2 Perform the continuation analysis of the end of the analyticity proof, Section 7.2.1. Calculate the asymptotics as $m \to 0$ or $m \to m_c$ (take into account the results of [PZ95], try to improve on them).

The behaviour of the self-similar solutions with $\beta < 0$ is a bit surprising since the profile keeps being shrunk in space at the same time that is shrunk in height. In order to understand qualitatively this fact of concentration near the origin we propose to the reader the following problem.

Exercise 7.3 For the solution of Theorem 7.3, prove that for every $p < \infty$ we have

$$\|u(t)\|_{L^p(B_R(0))} \sim (T - t)^\gamma$$

with $\gamma = \overline{\alpha}$ if $m \geq m_s$, while $\gamma = \overline{\alpha} - (n\overline{\beta}/p)$ for $m < m_s$. $R > 0$ is kept fixed.

Exercise 7.4 Calculate the spread rate of solutions of the PME and the FDE for $m \geq m_c$, by obtaining a formula for $\overline{x}(t)$ for the model solutions. Prove first that $\overline{x}(t)$ is finite for $m > n/(n+2) > m_c$.

8

Logarithmic diffusion in 2D and intermediate 1D range

As a transition from the range $0 < m < m_c$ of fast diffusions with extinction to the superfast range $m < 0$ where instantaneous extinction may take place, we study in this chapter two cases of fast diffusion that offer extinction tied to non-uniqueness for the Cauchy problem.

The first instance is the range $-1 < m \le 0$ in one space dimension. Here, a simple and quite complete analysis can be made of the phenomenon of non-uniqueness caused by the possibility of imposing an outgoing flux at infinity.

We devote much space to the second case, the logarithmic diffusion equation in the plane, which offers an important instance of loss mass which is tied this time to geometry. Subsection 8.2.2 presents a case of weak smoothing effects, and Subsection 8.2.3 contains a study of self-similarity and extinction rates for integrable non-maximal solutions. A new weak local smoothing effect is proved Section 8.3.

This chapter deals mainly with solutions which extinguish in finite time, and can be considered in that sense as a continuation of the preceding chapters on extinction. However, an important difference must be observed: the existence of different extinction rates was obtained in those chapters as a function of the different exponents m of the equation and of the class of initial data. It will be obtained here as a consequence of the possibility of controlling the behaviour of the solutions of the same equation with the same initial data through the flux conditions at infinity. In other words, as a consequence of the non-uniqueness of solutions of the Cauchy problem.

8.1 Intermediate range $-1 < m \le 0$ in $n = 1$

The rather radical property of instantaneous extinction that we will study in Chapter 9 as a peculiar feature of the range $m < 0$ of superfast diffusion does not apply to space dimension $n = 1$ in the range $-1 < m \le 0$, though other remarkable properties appear in this range. For convenience of reference, we will give to this case the name of intermediate 1D range.

We have already mentioned in Section 2.2.1 that this range behaves in some sense like a supercritical range (we have $m > (n-2)/n$ after all). Therefore, data in $L^1(\mathbf{R})$

produce solutions in $L^\infty(\mathbf{R}^n)$ for all $t > 0$ and the decay estimates based on comparison with ZKB solutions apply. Similar effects apply when $u_0 \in M^p(\mathbf{R})$ with exponent $p > 1$.

Non-uniqueness and Neumann data at infinity Maybe the main novelty is the lack of uniqueness. We present results obtained with our collaborators. According to [ERV88], there is existence and uniqueness of a class of maximal solutions; a maximal solution is characterized by the property of conservation of mass in time. But there is no uniqueness without that restriction, and actually [RV90] proves that this 1D problem admits a unique solution for every integrable initial function $u_0(s) \ge 0$ and every pair of flux data at infinity of the form

$$\lim_{x\to-\infty} (u^{m-1}u_x)(x,t) = g(t), \qquad \lim_{x\to\infty} (u^{m-1}u_x)(x,t) = -f(t), \tag{8.1}$$

on the condition that $f(t), g(t) \ge 0$ and are for instance bounded.[1] The physical explanation is as follows: diffusion is so fast that, in some sense, infinity lies at finite distance for the particles that form the continuous medium, and this makes it possible to impose so-called *Neumann data at infinity*, but only in the sense of *outgoing flux*.

Actually, the solution depends continuously on the data (u_0, f, g) in the L^1-norm and for two solutions u_1 and u_2 we have

$$\begin{aligned}\int (u_1(x,t) - u_2(x,t))_+\, dx &\le \int (u_{01}(x) - u_{02}(x))_+\, dx \\ &\quad + \int_0^t \{(f_1(\tau) - f_2(\tau))_+ + (g_1(\tau) - g_2(\tau))_+\}\, d\tau.\end{aligned}$$

We propose another way of looking at this non-uniqueness result: given a non-trivial initial function $u_0 \ge 0$, we can use the flux data f and g to *control* the extinction time of the solution at will: T is determined by the conservation law

$$\int_{\mathbf{R}} u_0(x)\, dx = \int_0^T (f(t) + g(t))\, dt + \int_{\mathbf{R}} u(x,T)\, dx. \tag{8.2}$$

Put $u(x,T) \equiv 0$ at extinction time.

For later use (in proving the weak smoothing effect of Subsection 8.2.2) we need to make precise the sense in which the flux data are taken. When f and g are bounded functions, this sense is as limits in x for a.e. t. But when f and g are continuous the sense is for every t. This is independent of the regularity of the initial data. Indeed, we have a priori estimates of the form

$$-\frac{C_1\, u}{t} \le u_t = (\log u)_{xx} \le \frac{C_2\, u}{t}.$$

[1] The boundedness restriction can be relaxed.

which imply upon integration that the limits are taken uniformly in the sense that, e.g., at the left end:

$$|(\log u)_x - g(t)| \leq \frac{C}{t} \int_{-\infty}^{x} u(x,t)\, dx \tag{8.3}$$

and this quantity is small (uniformly in $t \in (\tau, T)$) when $x \to -\infty$. Similarly for f as $x \to \infty$.

Non-uniqueness and integrability Imposing non-trivial flux data forces the solution to behave like $u(x,t) \sim c(t)\,|x|^{1/m}$ for $m < 0$ (resp. $u(x,t) \sim e^{-c|x|}$ for $m = 0$) for fixed t, and such behaviour makes the solution integrable in space. But we must remark that such a behaviour can be imposed independently at both ends, $x = -\infty$ and $x = \infty$. In this way, we may have solutions with outgoing flux at $x = -\infty$ and integrable profiles at this end, while u_0 may not be integrable near $x = \infty$ (say, $u_0 \sim c$) and consequently $u(t)$ is neither. The conclusion is that non-uniqueness is not restricted to integrable solutions. Moreover, we have shown in [RV95] how to obtain non-unique solutions by imposing outgoing fluxes for initial data which are not integrable at either end, for instance, for $u_0(x) \equiv 1$. An interesting and very explicit case of non-uniqueness of non-integrable solutions is reported in Exercise 8.6 and extends to several space dimensions.

But in some sense this result is not characteristic. If we restrict our attention to solutions that are not integrable at either end for all times, then existence and uniqueness hold true for initial data $u_0 \in L^1_{\rm loc}(\mathbf{R}^n)$, as proved in [ERV88, RV95]. Indeed, the theory loses its singular character.

We stop the study here because we want to devote our attention to the very interesting case of planar logarithmic diffusion. We will use the results of this section in the course of that study. Some other topics, like lack of uniform continuity, existence of large solutions, asymptotic behaviour and non-existence of solutions with changing sign are commented upon in Section 8.4. Local smoothing effects are discussed in Subsection 9.3.3 and the singular limit $m \to 1$ is discussed in Subsection 9.4.1.

8.2 Logarithmic diffusion in $n = 2$. Ricci flow

The fast diffusion equation in the limit case $m = 0$ in two space dimensions is a favourite case of our text for various reasons and we will devote much space to it. From the point of view of our systematic study, this is the 2D end-point of the critical line, so this section can be seen as a continuation of Section 5.6. The limit equation can be written as

$$u_t = \Delta \log(u), \tag{8.4}$$

hence the popular name of *logarithmic diffusion*. We refer to it as (P_0) by comparison with the PME / FDE with exponent $m \neq 0$ which is labelled (P_m).

However, the problem has a particular appeal because of is application to differential geometry, since it describes the evolution of surfaces by Ricci flow. More precisely, it represents the evolution of a conformally flat metric g by its Ricci curvature,

$$\frac{\partial}{\partial t} g_{ij} = -2\,\mathrm{Ric}_{ij} = -R\, g_{ij}, \tag{8.5}$$

where Ric is the Ricci tensor and R the scalar curvature. If g is given by the length expression $ds^2 = u\,(dx^2 + dy^2)$, we arrive at equation (8.4). This flow, proposed by R. Hamilton in [Ha88], is the equivalent of the Yamabe flow in two dimensions. We explain the geometry in Appendix III for the reader's convenience. We remark that, according to formula (AIII.7), what we usually call the mass of the solution (thinking in diffusion terms) becomes here the *total area* of the surface.

Coming to the smoothing question, we recall that the smoothing effect $L^p(\mathbf{R}^2)$-$L^\infty(\mathbf{R}^2)$ has been proved for all $p > 1$. Existence of solutions has been studied systematically by various authors since the mid 1990s, see references mentioned in Section 8.4. Thus, in [REV97] we prove that for every $u_0 \in L^1_{\rm loc}(\mathbf{R}^2)$ there is at least a solution $u \in C([0,\infty) : L^1_{\rm loc}(\mathbf{R}^2)$. In case $u_0 \in L^1(\mathbf{R}^2)$ such a solution is never unique, and a maximal solution can be selected that vanishes in finite time, while there are infinitely many non-maximal solutions, that must of course vanish in a finite time (cf. what happens in the intermediate 1D range).

We concentrate next on the behaviour of such solutions with integrable data for log-diffusion. We recall two useful tools. Given a solution $u(x,t)$ defined in $Q = \mathbf{R}^2 \times (0,T)$,

$$\widetilde{u}(x,t) = k u(k^2 x, t) \tag{8.6}$$

is also a solution. This is a special case of the scaling transformation that preserves mass. Besides, the inversion formula

$$\overline{u}(x,t) = \frac{1}{|x|^4}\, u\left(\frac{x}{|x|^2}, t\right) \tag{8.7}$$

transforms solutions of equation (8.4) defined for $x \neq 0$ into new solutions defined in the same space domain. This inversion must be carefully used, since the behaviour at the origin changes in such a way that most of times a singularity appears at $x = 0$. See below.

8.2.1 Integrable solutions. The mass loss phenomenon

Let us assume that $u_0 \in L^1(\mathbf{R}^n)$. We recall that solutions exist globally in time for $n = 1$, while for $n \geq 3$ no solutions exist with such integrable data (as we will explain in the next chapter, they vanish in zero time). We are thus in a borderline situation for $n = 2$. Actually, it is proved that solutions exist and present the following effect: while for $m > 0$ the law of conservation of mass holds,

$$\int u(x,t)\,dx = \int u_0(x)\,dx,$$

this property is lost in the limit $m = 0$, and mass is lost at a certain rate. The following precise result is proved.

Theorem 8.1 *For every $u_0 \in L^1(\mathbf{R}^2)$, with $u_0 \geq 0$ there exists a unique function $u \in C([0,T) : L^1(\mathbf{R}^2))$ which is a classical (C^∞ and positive) solution of (8.4) in Q_T and satisfies the mass constraint*

$$\int u(x,t)dx = \int u_0(x)dx - 4\pi\, t. \tag{8.8}$$

This solution is maximal among the solutions of the Cauchy problem for (P_0) *with these initial data, and exists for the time* $0 < t < T = \int_{\mathbf{R}^2} u_0(x)\,dx/4\pi$. *Moreover, the solution is obtained as the limit of positive solutions with initial data* $u_{0\varepsilon}(x) = u_0(x) + \varepsilon$ *as* $\varepsilon \to 0$.

The case of the mysterious number 4π It follows from the result that there is a striking difference between $m > 0$ and $m = 0$. It looks as if the minimal mass loss at $m = 0$, which amounts to 4π units per unit time, takes place in some mysterious way. We will give several clues to the mystery.

Explanation 1 There is a very simple explanation of the phenomenon for radially symmetric data and solutions, by transforming the equation into a fast diffusion equation in $n = 1$. This is the approach of [VER96], where this class of solutions is completely described. The transformation, already used in [Ki93], is given by

$$u(r,t) = r^{-2} v(s,t), \qquad r = e^s. \tag{8.9}$$

It is readily checked that v satisfies

$$v_t = (\log(v))_{ss}. \tag{8.10}$$

Note that $r > 0$ transforms into $0 < s < \infty$, and that

$$\int_{\mathbf{R}^2} u(x,t)\,dx = 2\pi \int_{\mathbf{R}} v(s,t)\,ds.$$

As we have just said, this 1D problem admits a unique solution for every initial integrable function $v_0(s) \geq 0$ and every pair of flux data at infinity of the form (8.1) on the condition that $f(t), g(t) \geq 0$ and are for instance bounded. The following relation between the fluxes of the two functions is obtained

$$r\,\partial_r(\log u(r,t)) = -2 + \partial_s(\log v(s,t)), \qquad r = e^t. \tag{8.11}$$

This allows us to explain the mass loss phenomenon. Indeed, the weak solutions of the 2D problem must have zero flux ar $r = 0$ (by regularity and symmetry), which means that after transformation into $n = 1$, the 1D equivalent has flux= -2 at $s = -\infty$. It follows that the transformed solutions $v(s,t)$ cannot be the maximal solutions of the 1D Cauchy problem (they would have zero flux and conserve mass, while ours lose the amount of $f + g$ units of mass per unit time, and this sum is at least 2). Using this approach, the following general result is proved, cf. [VER96].

Proposition 8.2 *For every rearranged* $u_0 \in L^1(\mathbf{R}^2)$ *and for every bounded function* $f(t) \geq 2$, *there exists a unique function* $u \in C([0,T) : L^1(\mathbf{R}^2))$, *which is a radially symmetric and classical solution of* (P_0) *in* Q_T *and satisfies the mass constraint*

$$\int u(x,t)dx = \int u_0(x)dx - 2\pi \int_0^t f(\tau)\,d\tau. \tag{8.12}$$

It exists as long as the integral in the left-hand side is positive. The case $f = 0$ *corresponds to the maximal solution of the Cauchy problem. In any case, the solution is bounded for all* $t > 0$.

Thus, not only are there infinitely many solutions for every fixed non-trivial initial function u_0, but also the extinction time can be controlled by means of the flux data f. Moreover, these solutions satisfy the asymptotic spatial behaviour

$$\lim_{r\to\infty} r\,\partial_r(\log u(r,t)) = -f(t) \tag{8.13}$$

for a.e. $t \in (0,T)$. It follows that

$$\log(1/u(r,t)) \sim f(t)\,\log(r) \geq 2\log r \quad \text{as } r\to\infty.$$

It is further proved that for maximal solutions

$$u(r,t) \geq \frac{C(t)}{r^2\log r^2}. \tag{8.14}$$

This estimate will be used to justify the geometrical explanation below.

Explanation 2 Let us now consider a particular situation that can be later used to understand the general case. We try to start the evolution from a Dirac mass. Indeed, we have discussed in Subsection 5.5.4 the limit of the ZKB solutions $U_m(x,t)$ when $m>0$ and $m\to 0$. The explicit formulas (5.35) have allowed us to see that

(i) $U_m(x,t) \to 0$ as $m \to 0$ uniformly in the set $t \geq M/4\pi$. This is extinction in quite explicit form.

(ii) Let Ω_R be the complement of $B_R(0)$. The limit $m\to 0$ of the mass of U_m contained in Ω_R amounts precisely to $4\pi t$ as long as $4\pi t \leq M$. It contains the 4π we were looking for. Summing up, we have

Lemma 8.3 *In the above situation we get*

$$\lim_{m\to 0}\int_{B_R(0)} U_m(x,t;M)\,dx = M - 4\pi t, \tag{8.15}$$

whenever $0<R<\infty$ *is fixed and* $t<M/4\pi$. *The limit is zero if* $t\geq M/4\pi$.

(iii) Moreover, it is easy to see then that for every $0<t<M/4\pi$ we have

$$\lim_{m\to 0} U_m(x,t;M) = (M-4\pi t)\delta(x).$$

This means a striking fact: the logarithmic equation (8.4) posed in $\mathbf{R}^2$ with data a Dirac mass $u_0(x) = M\,\delta(x)$, does not admit a weak solution in the usual sense, though it admits as a generalized solution the expression

$$U_{m=0}(x,t;M) = (M-4\pi t)_+\,\delta(x). \tag{8.16}$$

The mass defect, 4π units of mass per unit time, is explained as a mass that has disappeared at infinity by an effect of superfast diffusion.[2] Let us sum up the situation in more physical terms: the point mass trickles into the medium at a rate of 4π

[2]In functional terms, this defect is a manifestation of Fatou's theorem.

units per unit time; unfortunately, we cannot see that diffused mass somewhere in the plane $\mathbf{R}^2$ because the fast diffusion for small intensities carries it immediately to $x = \infty$.

Explanation 3 The general case Using this precise computation as a tool, we can attack the mass loss effect for general data $u_0 \in L^1(\mathbf{R}^2)$, $u_0 \geq 0$, by using the ideology of the worst case that permeates this book, and applying it in the limit $m \to 0$. More precisely, the description of the mass loss phenomenon is based on the possibility of comparing solutions with different diffusivities. Indeed, we want to take the limit of (P_m),

$$u_t = \Delta(u^m/m), \tag{8.17}$$

for $0 < m < 1$, precisely as $m \to 0$. The normalization factor $1/m$ is important here, since in the limit we find solutions of the logarithmic diffusion equation (P_0), i.e., (8.4).

Theorem 8.4 *The maximal solution of the Cauchy problem for* (P_0) *with* L^1 *initial data is the limit as* $m \to 0$ *of the solutions of the Cauchy problems* (P_m).

Proof (i) Let $u_0 \in L^1(\mathbf{R}^2)$, $u_0 \geq 0$, $M = \int u_0(x)\,dx$, and let $u_m(x,t)$ be the solution of equation (P_m) with $m > 0$. We know that all the solutions $u_m(x,t)$ are positive, C^∞ smooth functions defined for all $x \in \mathbf{R}^2$ and $t > 0$. We use the concentration comparison, Theorem 1.3, for $m > 0$ in a standard way and pass to the limit $m \to 0$ to get for every $t > 0$

$$\lim_{m\to 0} \int_{|x|\leq R} u_m^*(x,t)\,dx \leq \lim_{m\to 0} \int_{|x|\leq R} U_m(x,t)\,dx = M - 4\pi\, t.$$

We recall that u_m^* is the symmetrization of u_m with respect to the space variable.

(ii) We have to show that the converse inequality holds when $R = \infty$ and $0 < t < M/4\pi$. We proceed in several steps. In the first one, we assume that u_0 is radially symmetric, bounded and compactly supported. Then we recall the theory developed in [VER96] according to which equation (P_0) can be solved for these initial data with so-called Neumann data at infinity of the form

$$-\lim_{r\to\infty} \frac{r u'(r)}{u(r)} = a,$$

whenever $a \geq 2$. The mass loss of this solution is $2\pi a$ units of mass per unit time. This has been explained above.

But then we recall the possibility of comparing solutions of equations with different nonlinearities in Theorem 1.3 and notice the fact that the nonlinearity $\phi_0(s) = \log(s)$ is more diffusive than $C\,\phi_m(s) = C\,s^m/m$, $m > 0$, when acting on functions bounded above by a constant $L > 0$, if

$$\frac{1}{s} \geq C\,s^{m-1}, \quad 0 < s < L,$$

i.e., if $C \le L^{-m}$ (for the comparison of diffusivities see Subsection 1.1.5). Therefore, we may use the theorem to compare solutions with same initial data and different diffusivities and conclude that

$$\overline{u}(\cdot,t) \prec u_m(\cdot, C_m t),$$

where we denote by $\overline{u}(t) = \overline{u}(x,t)$ the solution of the logarithmic problem with flux data a. It follows that for all $R > 0$

$$\int_{|x|\le R} u_m(x,t)\,dx \ge \int_{|x|\le R} \overline{u}(x, C_m\,t)\,dx \ge M - a\,C_m\,t - \varepsilon(R),$$

where $\varepsilon(R) \to 0$ as $R \to \infty$. We may now pass to the limit and use the fact that $C_m \to 1$ as $m \to 0$ to conclude that the mass loss is exactly $4\pi\,t$ in the limit $m \to 0$.

(iii) The extension to radial functions $u_0(x)$ without the assumptions of boundedness or compact support is immediate by density. The case of radial solutions is solved.

(iv) The extension to non-radial solutions of the lower bound for the mass proceeds by comparison with radial solutions. Assume first that u_0 is continuous and positive. We find a radially symmetric function of mass $M' < M$ such that $0 \le u_0'(x) \le u_0(x)$. This can always be done, and we may even shift the origin of coordinates if we want to find a larger value of M'. Let u_m and u_m' the respective solutions, $u_m(x,t) \ge u_m'(x,t)$. By the L^1-contraction property we have

$$\int (u_m(x,t) - u_m'(x,t))\,dx \le M - M'.$$

Indeed, equality holds since they are ordered. On the other hand, we have proved that in the limit u_m' loses exactly $4\pi t$ units of mass if $0 < t < M'/4\pi$. In this interval we have then

$$\lim_{m\to 0}\int_{|x|\le R} u_m(x,t)\,dx \ge M - M' + M' - 4\pi t - \varepsilon(R) = M - 4\pi t - \varepsilon(R),$$

and the result is proved for $0 \le t \le M'/4\pi$. We continue by iteration until no mass is left, using the fact that solutions of equations (P_m), $m \ge 0$, are continuous and positive as long as they do not vanish, cf. [REV97]

(v) The extension to all functions $u_0 \in L^1(\mathbf{R}^n)$ is done by approximation using the L^1 contraction of the PME for all exponents m. We ask the reader to fill in the details. We also skip the justification of the properties of the limit solution which can be found in the previous reference. □

Explanation 4 The value $4\pi t$ has a very precise geometrical meaning, related to the Gauss–Bonnet formula. According to the formulas derived in Appendix III, if our constructed solution corresponds to the conformal factor of a complete two-dimensional surface, then it must lose exactly 2π times the Euler characteristic of M which should be 2, hence the total of 4π per unit time for the mass loss. See the appendix for more details on this intriguing topic. The fact that the metric is complete depends on checking that $x = \infty$ stays at an infinite distance for the metric, i.e., if

$$\int_\gamma u(x,t)\,dr = \infty$$

along any curve that joins a point $P = x_0$ to $x = \infty$, and for every fixed $t > 0$. This depends on the lower bound (8.14) for all solutions that we have obtained first in the radially symmetric case, and holds then for general data.

Explanation 5 Mass loss and the universal decay estimates This is not a proof, but a confirmation. If we consider the extinction results of Chapter 5, we find the universal estimate for the decay of the M^{p_*}-norm, formula (5.4), which is valid for $0 < m < m_c$ and $n > 2$. In view of the value of constant $k(m,n)$ given by (5.5), if we take the liberty of admitting real numbers for the dimension n and let $m \to 0$ and $n \to 2$ we find as limit the value 4π. More precisely, we have

$$\lim_{m\to 0, n\to 2} \frac{d}{dt} \sup_{r>0} (|B_r|^\sigma \int_{B_r} u(x,t)\,dx) \le -4\pi \tag{8.18}$$

with $\sigma = (n(1-m)-2)/n(1-m) \to 0$. This is pure *numerology* for the result we are discussing and relates it to a much more general extinction phenomenon. See comment in Section 5.8.

8.2.2 Weak smoothing effect

Theorem 5.1 shows the existence of the strong smoothing effect L^p–L^∞ for all $p > 1$ in the log-diffusion case, $n = 2$, with precise exponents and constants. However, the strong effect cannot be true for $p = 1$, since it would also imply an L^∞-estimate for the limit of the approximations of the Dirac delta as an initial mass, and this is impossible in view of the analysis performed above, reflected in formula (8.16). But the weak form of L^1–L^∞ effect is true.

Theorem 8.5 *Every solution of the logarithmic diffusion defined in* $Q = \mathbf{R}^2 \times (0,T)$ *corresponding to integrable data* $u_0 \ge 0$ *is bounded for all positive times, and the bound is uniform for* $t \ge \tau > 0$ *(for a given solution).*

Proof (i) We start with a simple but representative situation. Let us also assume that u_0 is integrable, radially symmetric, and decreasing in $r = |x|$. If u_0 is bounded, there is nothing to prove. The case $u_0 \in L^p(\mathbf{R}^2)$ for some $p > 1$ has already been settled. So we should assume that this does not happen.

We consider the maximal solution $u(r,t)$ of the Cauchy problem, presented in Theorem 8.2. It can be obtained from classical, bounded, and maximal solutions u_n with data $u_{0,n}(r) = \min\{u_0(r), n\}$. We have explained how to transform these solutions into solutions $v_n(s,t)$ of a one-dimensional problem. We translate into two dimensions the one-dimensional smoothing effect, cf. [ERV88, Prop. 2.6.iii], to get the bound

$$u_n(r,t) \le C\,\frac{M^2}{r^2 t}, \tag{8.19}$$

that gives a coarse estimate of the situation near $r = 0$. We thus conclude that the increasing limit $u(r, t)$, which is the maximal solution of the original Cauchy problem, exists and is a classical solution of the equation except possibly at $r = 0$, and it satisfies the previous estimate.

We still have to check that $u(0, t)$ is bounded above for every $0 < t < T$. For this we will need a finer growth estimate near $r = 0$. Here is the argument: if we look carefully at the one-dimensional solution $v(s, t)$, we see that it solves (8.10) and has flux -2 as $s \to -\infty$. These flux data are taken at least in an a.e. sense, cf. [RV90]; we have also pointed out in Section 8.1 that they are taken for all $0 < t < T$. Therefore, for all $\tau > 0$ we have

$$\lim_{s \to -\infty} (\log v)_s(s, \tau) = -2.$$

Translating back this result by means of (8.11) we get $r\, \partial_r(\log u)(r, \tau) = 0$ as $r \to 0$, so that

$$\frac{\log u}{\log r} \to 0 \qquad \text{as } r \to 0,$$

for $\tau > 0$, which is not sufficient, but at least implies that $u(x, \tau) \in L^p(\mathbf{R}^2)$ for all $p < \infty$. Now, the smoothing effect from L^p into L^∞ has already been proved in Theorem 5.1. Therefore, $u(t) \in L^\infty(\mathbf{R}^2)$ for all $t > \tau > 0$, i.e., for all $t > 0$.

(ii) If the data are not radially symmetric, we may use the symmetrization results of [Va04b], as described in Theorem 1.3. This technique allows us to conclude that the non-radial solution admits as an L^p upper bound the L^p-norm of the solution of the symmetrized problem ($1 \le p \le \infty$).

(iii) Finally, non-maximal solutions are obviously bounded above by the corresponding maximal solutions that have the same initial data. □

8.2.3 Integrable non-maximal solutions

In the case of maximal solutions, no self-similar asymptotic model was found. On the contrary, there are self-similar models for the solutions with outgoing flux larger than 4π. The most typical one is the separate-variable solution

$$u(x, t) = \frac{8(T - t)}{(1 + x^2)^2}. \tag{8.20}$$

It represents the evolution of a surface with total area $\int u(x, t)\, dx = 8\pi(T - t)$, hence a loss mass rate of 8π, i.e.,

$$\frac{d}{dt} \int u(x, t)\, dx = -8\pi, \tag{8.21}$$

as the reader will easily check. Note that this is twice the minimal rate. This solution is analogous in $n = 2$ to the minimal solutions of the Yamabe problem ($n > 2$, $m = m_s > 0$), described in (7.19), (7.39) (hint: put $\lambda = 1$ in the latter). Therefore, this solution represents the conformal shrinking of a flat metric representable by stereographic projection as a standard metric on a sphere in $\mathbf{R}^3$. We may say that the whole evolution represents the shrinking of a ball into a point with a linear rate, and this aspect was pointed out by R. Hamilton [Ha88]. Note finally that this solution is transformed into itself by the inversion formula (8.7).

More generally, we can consider Type-II self-similar solutions of the form

$$u(x,t) = (T-t)^{\alpha} f(x\,(T-t)^{\beta}), \tag{8.22}$$

with radially symmetric profiles $f > 0$ and $\alpha = 2\beta + 1$. They can be constructed by the phase-plane analysis proposed in Chapter 7, or by transformation into the one-dimensional problems. Actually, with the notation of Chapter 7, the relevant system (7.14) simplifies for $m = 0$, $n = 2$, into

$$\dot{X} = -(\alpha + \beta X)Z, \quad \dot{Z} = (2 + X)Z, \tag{8.23}$$

where $\dot{X} = dX/dr$, $\dot{Z} = dZ/dr$. Contrary to the case $m > 0$, this system can be easily analysed. The solution is given for $\beta \neq 0$ by the formula

$$\beta Z + X - \frac{1}{\beta}\log\left(1 + \frac{\beta}{2\beta+1}X\right) = 0. \tag{8.24}$$

The case $\beta = 0$, $\alpha = 1$, leads to the explicit solution above, which reads in this formulation $Z = -(1/2)X(X+4)$. Let us sum up the result.

Theorem 8.6 *For every $\alpha > 0$ there is a unique solution of problem* (8.29) *of the form* (8.22) *up to scaling. All these solutions are integrable and have a mass loss rate $M'(t) = -2\pi c$, with c a decreasing function of α.*

Scaling means transformation (8.6). We indicate the main steps of the proof of these facts in Exercise 9.3, and we ask the reader to work it out. It is proved that $c(\alpha) \to 2$ as $\alpha \to \infty$, $c(1) = 4$, and $c(\alpha) \to \infty$ as $\alpha \to 0$. A complete derivation of these profiles has been done [VER96, Appendix, page 42] though the notation is different and the conclusion is not looked at in the present form.

In this way, we parallel in some sense the results of Chapter 7 by finding a Type-II self-similar solution for all exponents $\alpha > 0$ (cf. with the results of Corollary 7.5). There are differences: now all the solutions are integrable in space, and there is no solution of this class with slowest space decay.

The phenomenon of relative concentration, described in Subsection 7.2.3 for the special self-similar solutions of Chapter 5 in the range $0 < m < m_s$, has a parallel here. When we look at the imposed flux rates, we find the following classification.

Proposition 8.7 *(i) For flux rates $-M'(t) \in (4\pi, 8\pi)$, the self-similar solutions* (8.22) *spread because $\beta > 0$.*

(ii) For $-M'(t) > 8\pi$ solutions concentrate towards Dirac deltas in relative size as $t \to T$, since $\beta < 0$.

(iii) For $M'(t) = -8\pi$ we have separation of variables and stabilization in relative size.

Also the maximal solution, i.e., the case $M'(t) = -4\pi$, spreads in time as we will see next, though there is no self-similar form.

Geometry again. Lifting the equation to the 2-sphere In the same way as the FDE with exponent m_s in dimension $n \geq 3$ was lifted in Subsection 7.5.2 to the sphere $\mathbb{S}^n$ by means of stereographic projection, cf. Theorem 7.9, we can do it here in dimension $n = 2$ with $m = 0$. The formula is

$$v(y, \tau) = \frac{1}{4}(1 + |x|^2)^2 u(x, t), \quad y = \Pi(x), \tag{8.25}$$

where $y = (\varphi, \theta)$ denotes a point in the sphere $\mathbb{S}^2$ and Π is the stereographic projection from $\mathbf{R}^2$ into $\mathbb{S}^2 \subset \mathbf{R}^3$ given by

$$\Pi(x_1, x_2) = \left(\frac{2(x_1, x_2)}{1 + |x|^2}, \frac{1 - |x|^2}{1 + |x|^2}\right). \tag{8.26}$$

This is an admissible change of framework in our analysis since Π is a conformal transformation (with precise conformal factor $4/(1 + x^2)^2$). The new time is given as usual by $\tau = \log(T/(T - t)) \in (0, \infty)$. Indeed, this approach was taken by Hamilton in 1988 [Ha88] and leads to the semilinear parabolic equation

$$v_\tau = \Delta_S \log v - 2, \tag{8.27}$$

where Δ_S denotes the Laplace–Beltrami operator in $\mathbb{S}^2$; the equation is posed for $(y, s) \in \mathbb{S}^2 \times (0, \infty)$. In principle, it is not clear whether the equation is satisfied at the point corresponding to $x = \infty$ (the North or the South Pole depending on the choice of projection, let us take the North Pole N); an easy analysis shows that it only makes good sense when the conditions at infinity are of the form

$$u(x, t) \sim C(t)|x|^{-4},$$

which implies that v will be bounded and then continuous and smooth by interior regularity for semilinear parabolic equations (local theory on the manifold). The absence of a singularity at that point is what characterizes the 'good geometrical flow', which according to what we have seen is the case when $M'(t) = -8\pi$. A unique smooth flow is constructed in $\mathbb{S}^2$. The most peculiar example, solution (8.20), transforms into a space-homogeneous function,

$$v(y, \tau) = 2(T - t),$$

which clearly represents a shrinking ball. But this solution admits variants that originate from space shifts of (8.20) that are written in the form

$$u(x, t; x_0) = \frac{8(T - t)}{(1 + (x - x_0)^2)^2}, \tag{8.28}$$

with $x_0 \in \mathbf{R}^2$ arbitrary. We ask the reader to write down the functions $v = v(y, \tau; x_0)$ of this family; they are not space-homogeneous but they are separable solutions and they represent shrinking balls anyway.

Other fluxes In the rest of the cases (in particular, all the other self-similar solutions with $M'(t) \neq -8\pi$), lifting the solution to the sphere can be done, but the equation is not satisfied by v at the corresponding pole N, and a standing singularity takes place there. This situation is described in technical terms as a *non-removable singularity*. Note that there are two options, $v(N) = \infty$ or $v(N) = 0$.

8.2.4 Asymptotic behaviour

We report here on the asymptotic in time behaviour of solutions of the Cauchy problem

$$\begin{cases} \partial u/\partial t = \Delta \log u & \text{in } \mathbf{R}^2 \times [0, T) \\ u(x, 0) = u_0(x) & \text{on } \mathbf{R}^2 \end{cases} \tag{8.29}$$

with $0 < T \leq \infty$ and initial data $u_0 \geq 0$ and locally integrable. Because of non-uniqueness, we must specify the class of solutions under consideration.

Maximal solutions We first consider the class of *maximal* solutions, which have interesting geometry and exhibit a delicate behaviour. The basic fact to be announced is: there is no self-similar model for the class of maximal integrable solutions of the log-diffusion equation in $n = 2$ (due to the failure of existence of solutions with a Dirac mass as initial data). J. King [Ki93] has formally analysed the extinction behaviour of maximal solutions u of (8.29) satisfying

$$\int_{\mathbf{R}^2} u(x, t)\, dx = \int_{\mathbf{R}^2} u_0(x)\, dx - 4\pi t$$

as $t \to T$, with $T = 1/4\pi \int_{\mathbf{R}^2} u_0(x)\, dx$. His formal asymptotics show that if u_0 is radially symmetric and compactly supported then, as $t \to T$, the maximal solution u enjoys the asymptotics

$$u(x, t) \approx \frac{2\,(T - t)^2}{T\,(r^2 + \lambda\, e^{2T/(T-t)})}, \qquad \text{for } r = O(e^{T/(T-t)}) \tag{8.30}$$

for some constant $\lambda > 0$, while

$$u(x, t) \approx \frac{2T}{r^2 \log^2 r}, \qquad \text{for } (T - t) \ln r > T. \tag{8.31}$$

Note that formula (8.30) means an exponentially small size of the solution as $t \to T$ (very slow extinction). The above (non-self-similar) asymptotics have not been shown rigorously. However, Daskalopoulos and Hamilton [DH04] have recently established sharp geometric estimates on the 'geometric width' and the 'maximal curvature' of maximal solutions of (8.29) near their extinction time for solutions of the logarithmic fast diffusion equation in the plane with bounded curvature. In this case, the area is given in terms of the maximal time $T < \infty$ by $A(t) = 4\pi(T - t)$. They prove there exists $C < \infty$ such that the width w and maximum curvature $R_{\max}$ satisfy for all $t \in (0, T)$

$$\frac{T - t}{C} \leq w\,(t) \leq C\,(T - t)$$

and

$$\frac{1}{C\,(T-t)^2} \le R_{\max}\,(t) \le \frac{C}{(T-t)^2}.$$

Integrable non-maximal solutions We consider next the extinction behaviour of non-maximal radially symmetric solutions of (8.29). Assume that $u_0 \ge 0$ is a radially symmetric, non-trivial function. Then, we know that for every $c > 2$ there exists a unique solution u of the log-diffusion equation (8.29) defined in $\mathbf{R}^2 \times (0, T)$ with $T = \int_{\mathbf{R}^2} u_0(x)\,dx/2\pi c$, and satisfying

$$\int_{\mathbf{R}^2} u(x,t)\,dx = \int_{\mathbf{R}^2} u_0(x)\,dx - 2\pi c\,t, \qquad 0 < t < T. \tag{8.32}$$

This solution is bounded for $t \ge \tau > 0$. It has the behaviour

$$\lim_{r\to+\infty} \frac{r\,u_r(r,t)}{u(r,t)} = -c, \quad \text{uniformly on } [a,b], \text{ as } r \to \infty \tag{8.33}$$

for any $0 < a < b < T$ (as shown in [ERV88]). S.Y. Hsu [Hs03, Hs04] studied the extinction problem and proved that there exist unique constants $\alpha > 0$, $\beta > -1/2$, $\alpha = 2\beta + 1$, depending on c, such that the rescaled function

$$v(y,\tau) = \frac{u(y/(T-t)^{\beta}, t)}{(T-t)^{\alpha}},$$

where $\tau = -\log(T-t)$, will converge uniformly on compact subsets of $\mathbf{R}^2$ to $\phi_{\lambda,\beta}(y)$, for some constant $\lambda > 0$, where $\phi_{\lambda,\beta}(y) = \phi_{\lambda,\beta}(r)$, $r = |y|$ is radially symmetric, smooth and positive and satisfies the ODE

$$\frac{1}{r}\left(\frac{r\phi'}{\phi}\right)' + \alpha\,\phi + \beta r\phi' = 0 \qquad \text{in } (0,\infty)$$

with data $\phi(0) = 1/\lambda$, $\phi'(0) = 0$. Actually, the relation between α and c is given by Theorem 8.6 and $\phi_{\lambda,\beta}(y)$ is just a scaling of the self-similar profile f of formula (8.22). Of course, in the case where $c = 4$ the above result simply gives the asymptotics

$$u(x,t) \approx \frac{8\lambda(T-t)}{(\lambda + |x|^2)^2} \qquad \text{as } t \to T$$

corresponding to the geometric result of Hamilton [Ha88] and B. Chow [Ch91] that, under the Ricci flow, a two-dimensional compact surface shrinks to a sphere. In that case it is worth working on the sphere $\mathbb{S}^2$ as we have explained at the end of Section 8.2.3, and even to renormalize the solution by introducing the function

$$w(y,\tau) = v(y,t)/(T-t), \tag{8.34}$$

and then w satisfies

$$w_\tau = \Delta_S \log w - 2 + w. \tag{8.35}$$

This result is more general in the sense that it does not need the assumption of symmetry. But then the possible asymptotic behaviours are given by the whole family of profiles $w(y, x_0)$ rescaled from the solutions $v(y, \tau; x_0)$.

8.2.5 Non-integrable solutions

If the initial data u_0 is a locally integrable function but $\int_{\mathbf{R}^2} u_0(x)\,dx$ is infinite, then there exists a solution u of the Cauchy problem (8.29) on $\mathbf{R}^2 \times [0, \infty)$. Uniqueness of such solutions is guaranteed in the radially symmetric case. More generally, uniqueness is proved in [REV97] under the assumption that the solution satisfies the asymptotic behaviour in space

$$\frac{1}{u(x,t)} = O(r^2 \log r^2/t) \quad \text{as } r = |x| \to \infty. \tag{8.36}$$

The gist of the condition is to impose a restriction of not decaying very fast at infinity, so that the effect of very fast diffusion will not take place in the form of non-trivial Neumann boundary conditions. See also Exercise 8.5.

It is not possible to ask for a uniqueness result for non-radial solutions with non-integrable data if we allow for fast decay along some range of directions. Thus, the non-unique solutions of the same equation in one space dimension, let us call them $v(x_1, t)$, with Neumann conditions at $x_1 \to \infty$ can be considered as solutions in two dimensions by putting

$$u(x_1, x_2, t) = v(x_1, t).$$

In this way, we can construct for the same non-integrable initial data an infinite collection of non-integrable solutions in $Q = \mathbf{R}^2 \times (0, T)$ that extinguish in a finite time T that can be prescribed. The question of uniqueness in this general setting is still partially understood. See also the very explicit example of non-uniqueness of Exercise 8.6.

Eternal solutions Let us also mention that the analysis of self-similarity produces the family of eternal solutions with infinite mass

$$u_{a,k}(x,t) = \frac{1}{a|x|^2 + k\, e^{4at}} \tag{8.37}$$

with $a, k > 0$. This is the version for $n = 2$ of the family that we have found in (5.36) for $m = m_c$ and $n \geq 3$.

Asymptotic behaviour This question is discussed in the literature under strong restrictions on the data. Here is an example: S.Y. Hsu [Hs04] studied the asymptotic profile of u under the assumption that the initial data u_0 satisfies the bounds

$$\frac{1}{a(|x|^2 + k_1)} \leq u_0 \leq \frac{1}{a(|x|^2 + k_2)} \tag{8.38}$$

for some constants $a > 0$, $k_1 > 0$, $k_2 > 0$. This makes it similar to the solution (8.37). Under such assumption on the data, the Cauchy problem (8.29) admits a unique solution u on $\mathbf{R}^2 \times [0, \infty)$ which satisfies the bounds

$$\frac{1}{a(|x|^2 + k_1 e^{4at})} \leq u(x,t) \leq \frac{1}{a(|x|^2 + k_2 e^{4at})}. \tag{8.39}$$

It is then shown in [Hs02b] (see also in [Wu93, Wu93b]) that the rescaled function

$$w(x,t) = e^{4at} u(e^{2at} x, t)$$

will converge, as $t \to \infty$, uniformly on $\mathbf{R}^2$ and also in $L^1(\mathbf{R}^n)$, to the function

$$\phi_{\beta,k_0}(x) = \frac{1}{a(|x|^2 + k_0)}$$

for some unique constant k_0 satisfying

$$\int_{\mathbf{R}^2} (u_0 - \phi_{a,k_0})\, dx = 0.$$

8.3 Weak local effect in log-diffusion

We report here on a very recent result obtained in [V05b] that extends the weak effect described in Subsection 8.2.2. Here is our improvement of that result in the form of a local effect.

Theorem 8.8 *Let $u(x,t) > 0$ be a smooth solution of the planar logarithmic diffusion equation defined in a bounded cylinder $Q = B_R \times (0,T)$ with initial data in $L^1_{\text{loc}}(B_R)$. Then for every $t_1, t_2 > 0$ and $R' < R$, u is uniformly bounded in $Q' = B_{R'} \times (t_1, t_2)$ with a bound that depends on R, R', t_1, t_2 and u_0.*

There are three cases in the proof: first and second, for radially symmetric functions, then for general functions.

1. Globally defined radial case. We just use the transformation into 1D to get a function $v(s,t)$, $s \in \mathbf{R}$, that evolves according to 1D log-diffusion with locally integrable data. Note that the origin $r = 0$ maps into $s = -\infty$. The result is true by some simple comparison away from the origin $R_0 < r < R$, i.e., for s away from $\pm\infty$. The strong local 1D effect was obtained in [ERV88] and will be reviewed in Subsection 9.3.3.

On the very end, $s = -\infty$, we have an estimate that has to be transformed into $r = 0$ as is done in the weak global effect in Theorem 8.5 above. In the first step it is proved that it belongs to L^p_{loc} for all p. Then, a more classical smoothing result is applied that implies that for data in the spaces $L^p(\mathbf{R}^2) + L^\infty(\mathbf{R}^2)$, $1 < p < \infty$, the solutions are bounded. See Theorem 8.5 above.

2. General radial case. (i) The last part of the previous argument works for globally defined solutions. In order to prove boundedness, including also $r = 0$, for solutions which are locally defined in space, we use as comparison function a certain $u_2(r, t)$ with the following properties: it is a radially symmetric solution of the PLDE defined in $Q_1 = B_R(0) \times (0, \infty)$, it is smooth and positive, it is increasing in the radial direction and has blow-up on the lateral boundary $\Sigma = \partial B_R \times (0, T)$. We can take a solution of the form

$$u_2(r, t) = (t + 1)\rho(r), \quad r = |x|. \tag{8.40}$$

The existence of such solutions is easy to conclude by solving the elliptic problem (ODE) satisfied by $\rho(r)$; it is proved in detail in Appendix A of [V05b]. We will use that the profile ρ blows up at $|x| \to R$ like

$$\rho(r) \sim \frac{2}{(R - r)^2}. \tag{8.41}$$

(ii) We take the quotient $w = u/u_2$. By virtue of the equations satisfied by u and u_2, we have

$$u_2 w_t + u_{2,t}(w - 1) = \Delta \log w. \tag{8.42}$$

In view of the behaviour of ρ, we know that $w \leq 1$ near the lateral boundary Σ. Actually, $w \to 0$ there.

Now, we introduce the function $z = \sup\{w, 1\} - 1 = (w - 1)_+$ we multiply the equation by $p(\log w)$, where $p(s)$ is a Brezis function approximating $\text{sign}_+(s)$, and we use Kato's inequality, $\Delta j(u) \geq j'(u)\Delta u$, which implies $\Delta u_+ \geq \text{sign}_+(u)\Delta u$, to get

$$u_2 z_t \leq \Delta \log(1 + z) - u_{2,t} z \leq \Delta \log(1 + z). \tag{8.43}$$

Recall that $z_t = w_t \text{sign}_+(w - 1)$ and that

$$\text{sign}_+(\log(w)) = \text{sign}_+(w - 1).$$

Inequality (8.43) describes our function z as a subsolution of a logarithmic diffusion equation in Q but for the presence of the *weight* u_2. Note also that z takes the value zero on the lateral boundary in a continuous way.

(iii) In order to get rid of u_2 and at the same time to work in the whole space, we can perform a stretching of variables that is taken from the study of symmetrization with weights of paper [ReV05]: if $x = (r, \theta)$ in standard polar coordinates, we take new polar coordinates $y = (s, \theta)$ such that $\rho(r) r dr = s ds$ for $0 < r < R$. This is a kind of mass change, as we will explain in more detail below. Note that as $r \to R$ we have $s \to \infty$ with the estimate

$$s(r) \sim (\log(R/r))^{-1/2} \sim (R - r)^{-1/2}$$

as $r \to R$. Note also that $s(r)/r \to (\rho(0))^{1/2} > 0$ as $r \to 0$. In this way we define a bijective map $x \mapsto y = T(x)$ from $B_R(0)$ onto $\mathbf{R}^2$. Changing the time variable into $\tau = \log(t+1)$ and expressing z as a function of (s, τ), $z(r,t) = Z(s, \tau)$ we have

$$Z_\tau \leq \frac{t+1}{u_2 r}\frac{d}{dr}\left(r\frac{d}{dr}(\log(1+z)\right) = \frac{d}{sds}\left(a(s)s\frac{d}{ds}(\log 1 + Z)\right) \tag{8.44}$$

The coefficient is given by $a(s) = r^2\rho(r)/s^2$. As $r \to R$ we have

$$a(s) \sim (R-r)^{-1} \to \infty.$$

We conclude that $a(s) \geq c$ for all $y \in \mathbf{R}_+$.

(iv) The standard symmetrization for parabolic equations allows us to compare z with the solution of the equation

$$Z_\tau = c\,\Delta_y(\log 1 + Z) \tag{8.45}$$

taking as initial data the spherical symmetrization of the function $Z(y,0) = Z(s,\theta,0) = z(x,0)$. Let us call this solution $\overline{Z}(y,t)$.

There are two ways of ending the argument. Function $Z(y,0)$ belongs to $L^1(\mathbf{R}^2)$ since we have $dy = s ds d\theta = \rho(r) r dr d\theta = \rho(r) dx$, hence

$$\int_0^\infty Z(y,0)\,dy = \int_0^R (w(x,0)-1)_+\,\rho(r)dx = \int_0^R (u(x,0)-u_2(r,0))_+\,dx \leq C.$$

In that case, as slight variation of the weak smoothing effect for solutions in the whole space with L^1-data proved above shows that $\overline{Z}(\cdot,\tau) \in L^\infty(\mathbf{R}^2)$ for all $t > 0$. The bound is of weak type, depends on some information on the initial function (not only the integral). As a consequence of the standard symmetrization result, we have

$$\|Z(\cdot,\tau)\|_\infty = \|z(\cdot,t)\|_\infty \leq \|\overline{Z}(\cdot,\tau)\|_\infty < \infty. \tag{8.46}$$

An alternative way is to take into account that we already know that if this latter function belongs to some space $L^p(\mathbf{R}^n)$ with $p > 1$, then we know that the strong smoothing effect $u(x,t_1) \in L^p_{loc}(\mathbf{R}^2)$ for some $p > 1$ by the local analysis done at $x = 0$. This means that $z(x,t_1) \in L^p(B_R)$, that $Z(y,\tau_1)$ is in $L^p(\mathbf{R}^2)$, and then the strong smoothing effect implies boundedness for $t > t_1$.

3. General non-radial data. If u is not radial, we use a new symmetrization argument. The theory of symmetrization for elliptic and parabolic equations which have weights like the factor $\rho(r)$ in (8.43) has been developed recently in a paper with [ReV05], where the equation is written as

$$\rho(x)u_t = L\varphi(u) + g(x,t),$$

L a linear uniformly parabolic operator, that we will assume here to be the Laplacian, and g is an integrable function. The equation can be replaced by an inequality like in (8.43), where $g = 0$ and $\varphi(s) = \log(s)$.

The application of the technique of symmetrization with respect to the measure ρ of [ReV05] uses the a transformation $y = T(x)$ exactly as described in the previous step. For reference, it is done in any dimension $d \geq 2$ in standard spherical coordinates written as before, in order to conserve the elements of volume, which are given by $dy = s^{d-1} ds\, d\Omega_{d-1}$ and $dx = r^{d-1} dr\, d\Omega_{d-1}$,[3] the function $s(r)$ is defined by the ODE

$$s^{d-1} \frac{ds}{dr} = \rho(r)\, r^{d-1}, \tag{8.47}$$

plus the initial condition $s(0) = 0$. This is what we did in step (iii) above, since $d = 2$ in our case. The possibility of applying symmetrization to non-radial solutions is subject to the extra condition

$$\frac{ds}{dr} \geq K\, \rho^{1/2} \tag{8.48}$$

for a certain constant $K > 0$. This is true in our case since $s'(r) = \rho(r) r / s(r)$, in view of the known behaviour of $\rho(r)$ and $s(r)$ for $r \approx 0$ and $r \to R$.

Under all those assumptions, we can use the symmetrization theory of [ReV05] to compare our function $z(x, t)$, subsolution of $u_2 z_t = \Delta \log z$ that vanishes on the boundary of its domain $B_R(0)$, with the solution $\overline{u}$ of the Cauchy problem

$$\begin{cases} \overline{u}_t = C\, \Delta_y (\log(1 + \overline{u})) \\ \overline{u}(y, 0) = z^*_{0,\mu} \end{cases} \tag{8.49}$$

posed in $\Omega^*_\mu = \mathbf{R}^2$ as spatial domain for the variable y. The constant C is specified in that paper but is unimportant here. Since

$$\int_{\mathbf{R}^2} z^*_{0,\mu}(y)\, dy = \int_{B_r(0)} z_0(x)\, \rho(x) dx < \infty$$

(note that z_0 vanishes in a neighbourhood of the boundary), the μ-symmetrization of z_0 is integrable in $\mathbf{R}^2$. Now, we know that for problem (8.49) the weak smoothing effect applies in its global form. In particular, we know that $\overline{u}(y, t) \in L^\infty(\mathbf{R}^2)$ for every $t > 0$. The symmetrization comparison ensures that $z(x, t)$ is also bounded. □

8.4 Comments and historical notes

Section 8.1. One-dimensional equation

Applications The one-dimensional log-diffusion equation appears to describe the fluid dynamical limit of Carleman's equations for the interaction of two types of particles

[3] $d\Omega_{d-1}$ is the surface element on the unit sphere $\mathbb{S}^{d-1}$; it will not appear later.

moving against each other, see [Cl57, Ku73, McK75, LiTo97, SV05], The equation also arises in the study of the expansion of a thermalized electron cloud [LH76]. It appears also as a model for long Van der Waals interactions in thin films of a fluid spreading on a solid surface, if certain nonlinear fourth-order effects are neglected, and then u denotes the height of the liquid film, cf. [LMR76, Ge85]. It has been mentioned to model convective diffusion in ion exchange kinetics in [HP58]. Chayes, Osher and Ralston study the equation for $n = 1$ and $m < 0$ in connection with self-organized criticality, modelling avalanches in sandpiles [ChOR93]. The equation is also used (in an equivalent form) to describe the cooling of a fireball produced by a strong explosion of a local gas [Me89] and the special case $\gamma = 1$ is studied in [MSS, KD01].

Range $m \le 0$ and pressure formulation The equation with $m < 0$ has been studied by Bertsch, Dal Passo, Luckhaus and Ughi [U86, DL87, BU90, BDU90, BDU92] for all $n \ge 1$ in a series of papers where the pressure version of the equation is used. The variable is $v = u^{m-1}$, and that leads to equation (5.38): $v_t = v\Delta v - \kappa|\nabla v|^2$, with $\kappa = 1/(1-m)$, see Section 5.7. They study in that context the definition of viscosity solution and the questions of uniqueness and positivity, that for u means boundedness. We recall the work of Bertsch and Ughi [BU90] for the subject of local boundedness. We have already explained in Section 3.8 the local boundedness effects for the FDE in the range $m_c < m < 1$, cf. [HP85].

Lack of continuity The range $-1 < m \le 0$ is singular in many respects. Thus, it is shown in [Va05] that bounded non-negative solutions need not be uniformly continuous in their domain of definition; the Harnack inequality fails; moreover, bounded discontinuous solutions can be produced as limits of perfectly classical solutions. Even the complete positivity in space that is a typical property of fast diffusion flows fails and we can have zero-set patterns that we have called *needles* in [Va05].

Self-similarity There exist self-similar solutions of the Cauchy problem in the case $-1 < m < 0$, $n = 1$, that have constant flux at infinity, and thus extinguish in finite time. These solutions have the form of separate variables.

There also exist self-similar solutions with variable flux function of the form

$$u(x,t) = (T-t)^{\alpha} f(x\,(T-t)^{\beta}), \tag{8.50}$$

where the exponent β is not zero ($(1-m)\alpha = 2\beta + 1$) .

Asymptotic behaviour The asymptotic behaviour of the solutions with outgoing flux $f(t) > 0$ near the extinction time is partially understood. It has been studied by Hsu in [Hs02] in the log-diffusion case $m = 0$, under some extra conditions on the data: if $u_0 \in L^1(\mathbf{R})$ is an even function and is monotone decreasing in $r \ge 0$, $u_0 \ge 0$ on $\mathbf{R}$. Moreover, the flux function $f \in C([0,\infty))$ is a monotone increasing function, and there is extinction in finite time, then the function

$$v(x,s) = \frac{u(x,t)}{T-t}, \quad s = -\log(T-t) \tag{8.51}$$

converges as $s \to \infty$ to the function

$$v_\lambda(x) = \frac{2}{\lambda \cosh^2(x/\sqrt{\lambda})} \tag{8.52}$$

uniformly on every compact subset of $\mathbf{R}$, with $\lambda = 4/f(T)^2$.

Local and large solutions Solutions of the 1D FDE need not be defined in the whole of $\mathbf{R}$. Interesting in that respect is the existence of so-called *large solutions*, i.e., solutions that are defined in a finite or semi-infinite space interval and take infinite boundary values on the boundary located at finite distance. Existence of such solutions is a well-known consequence of the existence of local estimates. Maybe the most famous solution in that respect in the one-dimensional case is

$$U_m(x,t) = c_m\, t^{1/(1-m)} x^{-2/(1-m)}, \tag{8.53}$$

which are just a modification of the special solutions (5.2) so much used in Chapter 5 in connection with estimates in Marcinkiewicz spaces. Note that now we take as domain of definition for the space variable either $I_1 = (0, \infty)$ or $I_2 = (-\infty, 0)$. Note that for $m = 0$ we have $U_0(x,t) = 2t/x^2$. We have studied such singular solutions for $m > m_c$, $m > 0$ in all dimensions in [ChV02].

Changing sign There is still a curious novelty associated to the range $m \le 0$ in $n = 1$: the equation does not admit any solutions with changing sign. This phenomenon has been explored by Rodríguez and Vázquez in [RV02]; it is moreover shown that when we approximate the problem by a sequence of diffusion equations $u_t = \phi_n(u)_{xx}$ for which a unique solution exists, then in the limit three options may happen: either the limit solution is positive everywhere and takes some 'projected initial data', or it is negative everywhere in a similar fashion, or it is identically zero. Finding the new initial data implies solving an associated elliptic problem. This property extends to dimensions $n \ge 1$.

Section 8.2 The singular parabolic equation $u_t = \Delta \log u$ arises in space dimension $n \ge 2$ in a number of models in liquid film dynamics, gas kinetics, and plasma diffusion. Thus, in [DBD96] the authors study the model problem of a thin colloidal film spread over a flat surface, and u represents the thickness.

The equation has been studied systematically by various authors since the 1990s, starting with the papers by Daskalopoulos and DelPino [DP95], DiBenedetto and Diller [DBD96] and Esteban, Rodriguez and Vázquez [VER96, REV97], where the existence of solutions with mass loss was detected and studied. The non-existence of fundamental solutions of the equation is discussed by Hui [Hu94].

The connection of the logarithmic diffusion with Ricci flow and two-dimensional geometry is explained for instance in [Ha88, Wu93, Wu93b], see also the remarks in [REV97]. The results about the loss of mass in the limit $m = 0$ are taken from [VER96,

REV97]. The present proof of the amount of mass loss performs a complete derivation of the phenomenon by mass comparison techniques, and was not included in that paper which contains an alternative proof.

The condition $\int_{\mathbf{R}^2} u_0(x)\,dx \geq 4\pi T$ of Theorem 8.1 can also be seen as a restriction on solvability of the Cauchy problem in the domain $Q = \mathbf{R}^2 \times (0, T)$, and this is for instance the approach of [DP95]. In fact, the condition is necessary and sufficient in the class of data $\{u_0 \in L^1(\mathbf{R}^2),\ u_0 \geq 0\}$. But we prefer to see the solution as existing globally in time in the weak sense, with the proviso that it undergoes extinction ($u \equiv 0$) for $t \geq T$. This latter approach is natural when we obtain the solutions by means of the standard approximations with global classical solutions, e.g., by taking data $u_{0\varepsilon} = u_0(x) + \varepsilon$.

In Explanation 4 we use the fact that the maximal solutions represent conformal metrics in the plane where the point $z = (x, y) =$ infinity lies at an infinite distance, but still the area is finite. We point out that on the contrary, the self-similar solutions of Subsection 8.2.3, with flux larger than 4π, are all compact, the point of infinity of $\mathbf{R}^2$ lies in the metric at a finite distance. We explain in [REV97] that this point is a singular point of the surface unless flux = 8π, where the self-similar solution represents a shrinking ball. There are many non-integrable solutions, i.e., surfaces of infinite area. One of them is called by Hamilton and coworkers the *cigar solution* because of its asymptotic form.

In Explanation 5 we have related the extinction phenomena for $m > 0$, $n > 2$, and $m = 0$, $n = 2$. There is a striking difference to be noted between both situations: while in the former case there is uniqueness of solutions for the Cauchy problem, so that the different extinction rates correspond to different initial data, in the log-diffusion case in two dimensions, the same initial data admit different solutions (even classical solutions), existing in principle for different times (but see Exercise 9.2).

There are many papers on the issue of existence, non-uniqueness and asymptotic behaviour for the logarithmic fast diffusion equation. Some of them have been quoted above.

Subsection 8.2.2 The weak smoothing effect has been proved for radial solutions in [VER96], thanks to the transformation into the one-dimensional problem.

We point out that this weak smoothing effect is much easier to prove than the critical-line effects of Chapter 6, cf. Theorem 6.7.

Subsection 8.2.3 In applying the Gauss–Bonnet formula to the solutions (8.20) we recall that the Euler characteristic of the sphere is 2.

Geometrically, all the Type-II self-similar solutions for $\alpha \neq 1$ represent compact surfaces with a singular point corresponding to $|x| \to \infty$, see more on that issue in [VER96].

From the point of view of functional analysis the study of these solutions offers a simplified version of the analysis of self-similarity performed in Chapter 7. Actually, we are here sitting in the corner $m = 0$, $n = 2$ of the curvilinear triangle $\mathcal{T} = \{(m, n) : 0 < mn < n - 2\}$ considered in that chapter, and our analysis is a kind of germ of what happens inside $\mathcal{T}$. Note that we replace the moving parameters (m, n) by the new parameter $M'(t)$, or the pointwise flux $c = M'(t)/2\pi$.

Subsection 8.2.4 Many problems are still open in the area of asymptotics of logarithmic diffusion.

Open problem The extinction behaviour of *non-radial* solutions to (8.29) satisfying (8.32) with $c > 0$ is still an open question in the general case.

Subsection 8.2.5 The question of uniqueness is important and is often discussed in the literature, usually in a partial way. See, e.g., [Wu93, Hs04].

Section 8.3 As we have said, the L^∞ local bound must depend on the initial data through some information beyond the L^1-norm, otherwise it would be true in the limit for Dirac delta as initial data (and it is not, as we show in all detail). Inspection of the proof shows that the dependence on the initial data takes place only trough the concentration properties of the symmetrized function of u_0/ρ, in particular, the measure of its level sets. This is stable under approximation.

Log-diffusion for $n > 2$ The question of solvability of the log-diffusion equation is considered in dimensions $n > 2$ in [DP99]. They show that if $N(f)$ is defined by

$$N(f)(r) = \int_0^r \frac{1}{\omega_n s^{n-1}} \int_{B_s(0)} f(x)dx,$$

then there is no solution of the Cauchy problem if

$$\int_0^\infty \exp\left(-N(f)(r)/\tau\right) r dr = +\infty \tag{8.54}$$

for some $\tau \in (0, T)$. Furthermore, if f is radially symmetric, there exists a solution of the Cauchy problem posed in $\mathbf{R}^n \times (0, T)$ if and only if

$$\int_0^\infty \exp\left(-N(f)(r)/\tau\right) r dr < \infty \tag{8.55}$$

for all $\tau \in (0, T)$.

If $n = 2$, (8.55) reduces to the standard condition $\int u_0(x)\,dx \geq 4\pi T$ regardless of whether f is radially symmetric or not. This is far from true in dimensions $n \geq 3$ for non-radial f. The authors construct an explicit example for $n \geq 3$ to show this point.

A reason for the anomaly is the fact that solutions in $\mathbf{R}^n$ for $n > 2$ but depending only on the two first coordinates will exist precisely under the conditions for existence in $n = 2$.

The three-dimensional Ricci flow This has been proposed by Hamilton [Ha88] as a tool to prove the famous Poincaré conjecture, and is now being much discussed because of the achievements of Perelman, cf. [An04, Per03]. But let us warn the reader that the three-dimensional version is a blow-up problem, it is much harder than our present case and falls outside the scope of this text.

Exercises

Exercise 8.1 (i) Prove the assertion that transformation (8.9) maps the radial two-dimensional problem into a one-dimensional one.

(ii) Obtain the relation of fluxes.

Exercise 8.2 Using the result of Proposition 8.2, obtain two different radial solutions of the log-diffusion equation in $n = 2$ with same initial data u_0 and same existence time T.

Hint: control the flux function f.

Exercise 8.3 Construct the Type-II self-similar solutions with $\alpha > 0$ announced in Subsection 8.2.3 as follows.

(i) Study system (8.23) and prove formula (8.24). Show that Z is a concave function of X and that $Z'(0) = -2/\alpha$.

(ii) Show that the limit $\beta \to 0$ gives the solution (8.20).

(iii) Show that solution (8.20) is the limit of the solutions of the Yamabe case $m = m_c$ when we take the distinguished limit $m \to 0, n \to 2$ with $(n-2)/m \to 4$.

(iv) Show that for $-1/2 < \beta < \infty$, equation (8.24) admits a unique $X_* \neq 0$ such that $Z = 0$, and such X_* is negative. Plot the graphs of the profiles of the solutions in the (X, Z) plane as α varies. Show that in all cases $X_* < -2$ and $\beta X_* + 2\beta + 1 > 0$.

(v) More precisely, write $y = \beta X$ and show that

$$y = \log\left(1 + \frac{y}{2\beta + 1}\right)$$

has a unique solution. In case $\beta > 0$ show that a unique solution $y_* \neq 0$ exists and is negative and decreases from 0 to infinity as β grows; when $\beta \in (-1/2, 0)$, existence and uniqueness also happens but now $y_* > 0$ and y_* goes from 0 to ∞ as β decreases to $1/2$.

(vi) Show the asymptotic behaviour $X_*(\alpha) \approx \frac{1}{2}\log(\alpha)$ when $\alpha \to 0$, and $X_*(\alpha) \to -2$ as $\alpha \to \infty$ with $-2-(1/\beta) < X_* < -2$. Show also that $X_* = -4+(4/3)(\alpha-1) = -4 + (8/3)\beta$ for $\beta \approx 0$.

(vii) In order to derive the monotone dependence of X_* on α (or β), analyse separately the cases $\beta > 0$ and $\beta < 0$ in the above formula. In the latter case, show that $-X_*$ grows from 4 to ∞ as β decreases from 0 to $-1/2$.

(viii) For $\beta > 0$ you may argue as follows: if $F(X) := \beta X - \log(1 + \frac{\beta}{2\beta+1}X)$, we have

$$\frac{\partial F}{\partial X} = \beta^2 \frac{X+2}{2\beta X + 2\beta + 1}, \qquad \frac{\partial F}{\partial \beta} = \beta X \frac{2\beta(X+2) + X + 4}{(2\beta+1)(\beta X + 2\beta + 1)}$$

so that on the line $X = X_*(\beta)$ where $F(X_*, \beta) = 0$ we have

$$\frac{dX}{\partial\beta}=-\frac{\partial F/\partial\beta}{\partial F/\partial X}=-\frac{X(2\beta(X+2)+X+4)}{\beta(2\beta+1)(X+2)}.$$

We know that $X/(X+2)>0$, $2\beta+1>0$, so that X_* is an increasing function of β if we prove that $(X+2)(2\beta+1)+2=\alpha(X+2)+2<0$ along the solution curve. Note that it is true for $\beta\approx 0$. Show that this happens for all α by analysing the consequence of having a non-monotone function $X_*(\alpha)$.

(ix) Show that the decay rate of the solution $u(x,t)$ is $M'(t)=2\pi X_*$. Hence, $c=-X_*$.

Exercise 8.4 Prove Proposition 8.7. Show that the mean deviation is finite only for $M'(t)<-8\pi$ and calculate its evolution. Show in this case the precise statement about asymptotic convergence towards a Dirac delta.

Exercise 8.5 Check the existence of separable solutions for the 2D logarithmic diffusion equation. Show that there is one such solution of the form

$$U(x,t)=\frac{2t}{r^2(\log r)^2} \tag{8.56}$$

with $r=|x|\in(0,1)$. Find this solution by transformation of the 1D solution (8.53). Prove that it is self-invertible according to formula (8.7), but then the definition of the transformed formula works for $1<r$. Show that it is integrable at $x=0$ but not near $|x|=1$.

Exercise 8.6 A CURIOUS CASE OF NON-UNIQUENESS. The logarithmic diffusion equation admits in dimension $d=1$ the following family of stationary solutions

$$u(x,t)=Ae^{ax},\quad A>0,\, a\in\mathbf{R}. \tag{8.57}$$

When $a>0$ we observe that the flux at $x=\pm\infty$ is given by

$$f(t)=\lim_{x\to\pm\infty}\frac{u_x}{u}=a.$$

Since this quantity is non-zero and the solution is integrable at $x=-\infty$, there should be according to the theory another solution with flux zero at this end (the maximal solution). The existence of such a solution proves that the explicit steady state is not the natural solution provided by the theory. We propose in this exercise the method of construction in semi-explicit form.

(i) Let us put for definiteness, $A=a=1$. Try for the sought solution the form

$$u(x,t)=t\,f(x-\log t). \tag{8.58}$$

Prove that $f(s)$ must satisfy the equation

$$(f'/f)' = f - f'. \tag{8.59}$$

(ii) Prove that if we want the initial data to be e^x then we need $f(s)/e^s \to 1$ as $s \to \infty$. The condition of zero limit and zero flux at $x = -\infty$ is equivalent to $f(s),\ f'(s)/f(s) \to 0$ as $s \to -\infty$.

(iii) Introduce the function $g = f'/f$ and derive the equation

$$\frac{dg}{df} = \frac{1-g}{g}, \tag{8.60}$$

which upon integration gives for the maximal solution

$$f + g + \log(1-g) = 0. \tag{8.61}$$

(iv) Prove that $\log f(s)$ behaves as $s \to \infty$ like $s + O(e^{-e^s})$ while as $s \to -\infty$ we have

$$f(s) \sim \frac{2}{s^2}. \tag{8.62}$$

(v) Prove that $u_t = f - f' = f(1-g) > 0$ so that $u(x,t) > e^x$ for all $t > 0$ and all $x \in \mathbf{R}$. Calculate the approximate asymptotics of $u(x_0, t)$ as $t \to \infty$ in the form

$$u(x_0, t) \sim \frac{t}{(\log t)^2}. \tag{8.63}$$

(vi) Extend the non-uniqueness result to $n > 1$ by considering the solutions

$$U_1(x,t) = e^{x_1}, \quad U_2(x,t) = u(x_1, t).$$

9
Superfast FDE

In this chapter we address the question of extending the theory and estimates of the FDE to cover the singular range $m \le 0$, that has been only very partially considered in previous chapters. In this way we want to complete the scope of our investigation about decay, smoothing effects, and best constants. But the range offers us the possibility of learning about a novel and quite interesting dynamical issue: instantaneous extinction, which is tied to boundary layers at the initial time.

In the second part of the chapter we discuss the question of local estimates improving on results of previous sections. We pose an open problem.

9.1 Preliminaries

We have already modified the form of the equation to adapt it to this range as a parabolic equation, at least in a formal sense, see Subsection 1.2.2. The theory of solutions for the FDE in this range has been discussed in the literature, specially for the initial value problem posed in the whole space, and it offers interesting novelties. As we have already mentioned, a main question in this range is existence.

Let us recall the information we already have about the questions related to smoothing estimates. We have seen in Subsection 2.2.1 that in dimension $n = 1$ we may cover the range $m > -1$ in the standard theory, represented by ZKB solutions with L^1–L^∞ effects. We have also seen that the limit exponent in dimension $n = 2$ is $m = 0$ with its curious phenomenon of mass loss.

Moreover, on top of the critical line, i.e., for $2p > n(1 - m)$, we have standard existence and smoothing effects, even if $m \le 0$, as announced in Theorem 3.4. Actually, a main point of the theory is proving that solutions exist generally for data in such L^p or M^p spaces. The self-similar solutions associated to those smoothing effects are examples of solutions whose data go to zero as $x \to \infty$, but exist globally in time.

As for extinction, the explicit extinction example (5.2) exists for $m = 0$ and has the form

$$U(x, t; T) = \frac{A - 2(n - 2)t}{|x|^2}, \quad A > 0, \tag{9.1}$$

with existence time $T = A/2(n - 2)$. There is no problem in proving the extinction result like Theorem 5.2 corresponding to this solution. For $m < 0$ the explicit solution also exists and the same result holds. The value

$$p_* = n(1 - m)/2, \tag{9.2}$$

is now larger than $n/2$, that is all.

We also recall that the question of delayed boundedness discussed in Chapter 6 applies in this range for data in the borderline space $M^{p_*}(\mathbf{R}^n)$.

Important novelties appear on and below the critical line, and will be examined next.

9.2 Instantaneous extinction

Let us start by stating that the most important effect for $m \le 0$ is in our opinion the non-existence of solutions for data in spaces below the critical line. A general non-existence result was first proved in [Va92]:

Theorem 9.1 *No solutions exist with data in $L^1(\mathbf{R}^n)$ if $m < 0$ in dimensions $n \ge 2$. The same holds for $m = 0$ if $n \ge 3$. The non-existence range for $n = 1$ is $m \le -1$.*

According to the proof of [Va92], the absence of solution can be explained as follows. When the problem is approximated by solutions u_ε with data $u_0(x) + \varepsilon$ for $\varepsilon > 0$ (there is no problem for the existence of such solutions since the equation is not singular parabolic in that range of values), then as $\varepsilon \to 0$ the sequence u_ε converges uniformly to zero in sets of the form $\mathbf{R}^n \times (\tau, \infty)$ with $\tau > 0$. A discontinuity layer occurs therefore at $t = 0+$, allowing the initial data not to be taken in the limit. From the point of view of the present text, this phenomenon can be viewed like a smoothing effect with constant $c = 0$, or, better, as extinction in time $T = 0$, in other words, instantaneous extinction (IE).

Previous work in one space dimension had shown non-existence of solutions with L^1 data for $m \le -1$, cf. Herrero [H89], see also [ERV88]. On the other hand, for $p > 1$ the question of existence versus instantaneous extinction was completely solved when $m < 0$ by a result of Daskalopoulos and Del Pino [DP97], which give necessary conditions for solutions to exist or for solutions to have instantaneous extinction. They prove

Theorem 9.2 *Assume that $m < 0$, that $u_0 \in L^1_{loc}(\mathbf{R}^n)$, and that*

$$\lim_{R\to\infty} R^{n-(2/(1-m))} \int_{B_R} u_0(x)\,dx = 0.$$

Then, there is no solution in any time interval. In particular, this happens when $u_0 \in L^p(\mathbf{R}^n)$ with $p \le p_*$.

Remarks

(1) When $n = 1$ this condition needs $m < -1$. In fact, solutions with L^1 data exist for $m > -1$. They are not unique, but the maximal ones converge to the ZKB profiles, as shown in [ERV88, RV90].

(2) We remark that [DP97] uses a somewhat weaker integrated condition that we do not need to recall at this point. In any case, the existence situation borders optimality: solutions exist (and vanish in finite time), as we know, for some explicit data $u_0 \in M^{p_*}(\mathbf{R}^n)$, though they do not exist for $u_0 \in L^{p_*}(\mathbf{R}^n)$.

(3) The non-existence result of Theorem 9.2 is extended to the case $m = 0, n \geq 3$ in [DP99].

9.2.1 Self-similar approach to instantaneous extinction

The phenomenon of IE for $m \leq 0$ when the data belong to the Marcinkiewicz spaces $M^p(\mathbf{R}^n)$ with $p < p_*$ can also be established by using the same comparison and worst-case strategy of preceding sections. Since the result confirms the results of [DP97, DP99], we will only consider the case $m = 0$ and $n \geq 3$, and will also be rather sketchy. Our interest is directed to showing the way the phase-plane analysis can be used in this situation.

(i) The idea is to take the limit of the profiles corresponding to self-similar solutions with data

$$U_0(x) = A\,|x|^{-\gamma}, \tag{9.3}$$

with $A > 0$ and $\gamma = n/p$ and let $m \to 0$ with γ fixed. Our value of p lies in the interval $(1, p_*(0) = 2)$, hence $2 < \gamma < n$. For $m > 0$ small enough we have $2/(1-m) < \gamma < n < (n-2)/m$. We may now repeat the analysis of Subsection 5.4.1 and obtain solutions of the form

$$U(x,t) = t^a f_m(|x|\,t^b), \tag{9.4}$$

and similarity exponents

$$b = b(m) = \frac{1}{\gamma(1-m)-2}, \qquad a = a(m) = \frac{\gamma}{\gamma(1-m)-2}. \tag{9.5}$$

Introducing the variables

$$\xi = e^r, \qquad X(r) = \frac{\xi f'}{f}, \qquad \text{and} \qquad Y(r) = \xi^2 f^{1-m}. \tag{9.6}$$

we arrive at the system:

$$\begin{cases} \dot{X} = (2-n)X - mX^2 + b(m)(\gamma + X)Y \\ \dot{Y} = (2 + (1-m)X)Y \end{cases} \tag{9.7}$$

where $\dot{X} = dX/dr$. The limit system for $m = 0$ is

$$\begin{cases} \dot{X} = (2-n)X + b(\gamma + X)Y \\ \dot{Y} = (2 + (1-m)X)Y \end{cases} \tag{9.8}$$

with $b = b(0) = 1/(\gamma - 2)$.

As the profile is positive, we only need to consider orbits where $\{Y > 0\}$. Since it is monotone decreasing, we have $X < 0$. All together, we work in the quadrant $Q = \{X < 0,\ Y > 0\}$. As for end values, as $\xi \to 0$, i.e., $r \to -\infty$, we will have $f_m(\xi) \sim A\,\xi^{-\gamma}$, hence

$$X(-\infty) = -\gamma, \quad Y(r) \sim \xi^{-\gamma(1-m)+2} \to \infty$$

as $r \to -\infty$. Moreover, we have

$$X + \gamma \sim \frac{\gamma(n-2+\gamma)}{bY} = cY^{-1} \sim \xi^{\gamma(1-m)+2} \to 0.$$

Note that this behaviour is common as initial behaviour for all trajectories in Q. There is an infinity of them. The difference between them comes in the form of higher order terms.

(ii) We are interested in knowing what happens as $m \to 0$ with the connection

$$Y = Y_m(X), \qquad -(n-2)/m < X < -\gamma,$$

going from the asymptote $X = -\gamma$, $Y = +\infty$ to the point $(-(n-2)/m, 0)$ as r goes from $-\infty$ to ∞. Clearly, the point goes to the end of the horizontal axis, $Y = 0$, $X \to -\infty$. The curves are smooth and monotone increasing, and the asymptote is fixed. Note that all the curves lie on top of the respective vertical isocline, i.e.,

$$Y_m(X) > Y_{\infty,m} = \frac{(n-2-mX)X}{b(m)(\gamma + X)}.$$

We contend that the connection disappears as $m \to 0$. Indeed, in the limit all trajectories revolve in Q and slide along the X axis with a positive direction, to finish in the first quadrant with an asymptotic behaviour that means blow-up at finite distance for the actual solution. The reader can consult this type of argument in [FV03]. Therefore, we conclude that

$$\lim_{m \to 0} Y_m(X) = +\infty,$$

uniformly in compact subsets of $(-\infty, -\gamma)$.

(iii) The next step is to show that $f_m(\xi)$ goes to zero as $m \to 0$ uniformly for $|\xi| \geq \varepsilon > 0$.

Proof It is a technically delicate but essentially simple argument on the continuous behaviour of the trajectories emanating from the asymptote $X = -\gamma$ at $m = 0$. We parametrize this family of curves in the form $(X_{0,k}(r), Y_{0,k}(r))$ with a parameter $k \in \mathbf{R}$ such that as $k \in \mathbf{R}$ goes to ∞ the curves tend to go far into the region $X \sim -\infty$ before returning and blowing up at different points $X > 0$. Consequently, once we fix the behaviour as $r \to -\infty$

$$f(r) \sim Ae^{-\gamma r},$$

and at plus infinity,

$$X \to X_1,$$

we have a fixed orbit and $k \to \infty$ as $X_1 \to \infty$. We conclude that these curves satisfy $f_{0,k}(r) \to 0$ as $k \to \infty$ uniformly on compacts $[-L, L] \in \mathbf{R}$.

Next, we make a continuation argument to prove that for every $L = 1/\varepsilon > 0$ there exists m_ε such that for $0 < m < m_\varepsilon$ and $k \geq k_\varepsilon$

$$f_{m,k} \leq \varepsilon, \qquad r \in [-L, L].$$

Finally, for all k an easy comparison shows that

$$f_m(\xi) \le f_{m,k}(\xi).$$

Since $f_{m,k}$ is decreasing, we have $f_{m,k} \le \varepsilon, r \ge -L$. Thus, we conclude that the $L^1_{\rm loc}$ norm goes to zero. This means that the limit solution is zero in the $L^1_{\rm loc}$ sense.

(iv) Comparison by concentration means that when u is a bounded solution in M^p, $p < p_*$ and $m \le 0$, then it follows that the solution is necessarily zero.

(v) Continuity of the solutions with respect to the data in the $L^1(\mathbf{R}^n)$ norm allows us to eliminate the $L^\infty(\mathbf{R}^n)$ assumption. □

Indeed, the functions $f_{0,k}$ give a complete proof of non-existence by comparison, using the solutions that are lifted by ε at the beginning. For the conventional proof see the author's paper [Va92].

9.3 The critical line. Local smoothing effects

In Chapter 5 we have shown that there is no M^{p_*}–L^∞ effect for $m < m_c$ by showing an explicit solution that does not improve its space. This seems to settle the question of smoothing effect on the critical line, but it does not completely. We may ask the following question

Problem 9.1 Does $u_0 \in L^{p_*}_{\rm loc}(\mathbf{R}^n)$ imply that the solution becomes locally bounded for $t > 0$ when $m < m_c$?

Our guess is that the answer is yes. We have seen in Section 5.5 that the problem of extinction of solutions defined in $\mathbf{R}^n$ is a question about the behaviour of the solutions for large values of x, and has in principle nothing to do with the behaviour for large values of u that is what is of concern for smoothing effects.

There are a number of partial results in the direction of the conjecture:

(i) The answer to this local effect is positive when u_0 satisfies a more stringent condition, $u_0 \in L^{p_*}(\mathbf{R}^n) + L^\infty(\mathbf{R}^n)$, see Chapter 6.

Note that when $m < 0$ and $n \ge 2$, the results of Daskalopoulos and Del Pino [DP97, DP99] mentioned before show that the solution vanishes identically when $u_0 \in L^{p_*}(\mathbf{R}^n)$. A fortiori, the smoothing effect into any $L^\infty(\mathbf{R}^n)$ is true.

(ii) We give next a proof of the complete result in one dimension, using as technical tool the Bäcklund transform.

9.3.1 *One-dimensional analysis on the critical line*

We consider next the question of obtaining locally bounded solutions under conditions of local boundedness of integral norms on the data when $m < -1$. This local smoothing effect can be demonstrated in (and on top of) the whole critical line in one dimension. Therefore, only the case $p = (1-m)/2$ need be considered.[1]

[1] The reader will notice that this section is a continuation and improvement of Chapter 6.

We introduce the space $UL^p_{\mathrm{loc}}(\mathbf{R}^n)$ of uniformly locally L^p functions, i.e., functions $f \in L^p_{\mathrm{loc}}(\mathbf{R}^n)$, such that there is a radius R and a constant C satisfying

$$\int_{B_R(x)} |f(x)|^p\,dx \le C$$

uniformly in $x \in \mathbf{R}^n$. Clearly, $L^p(\mathbf{R}^n) + L^\infty(\mathbf{R}^n) \subset UL^p_{loc}(\mathbf{R}^n)$. Then we have

Theorem 9.3 *The following results are true when $n = 1$, $m \le -1$ and $p = (1-m)/2$:*

(i) If $u_0 \in L^p_{\mathrm{loc}}(\mathbf{R})$, then $u(t)$ is locally bounded in $Q = \mathbf{R} \times (0, \infty)$.

(ii) A weak smoothing effect is true for the one-dimensional FDE in that range, going from $UL^p_{\mathrm{loc}}(\mathbf{R})$ into $L^\infty(\mathbf{R})$.

(iii) If $u_0 \in L^p(\mathbf{R})$, then $u \equiv 0$ in $Q = \mathbf{R} \times (0, \infty)$.

Proof (i) By adding the initial data $u_0(x)$ with its reflection around a point $a \in \mathbf{R}$ we can find a larger data

$$\tilde{u}_0(x) = u_0(x) + u_0(2a - x),$$

that is symmetric around $x = a$. By uniqueness, so is the solution with respect to the space variable. We may also assume that $u_0 \ge \varepsilon > 0$ by adding a constant if needed. It will be enough to prove the result for the new data.

Our main technical tool is the Bäcklund transform (BT). It has the form of a non-local change of variables:

$$U = \frac{1}{u}, \quad y(x,t) = \int_a^x u(s,t)\,ds. \tag{9.9}$$

It is known that it transforms the problem into the conjugate equation

$$U_t = (U^{m'}/m')_{yy}, \tag{9.10}$$

where $m' = -m > 1$ if $m < -1$. This is a porous medium equation. $U(y,t)$ is symmetric around $y = 0$ and finite everywhere, hence continuous. The question is to decide whether $U(0,t) = 0$ for some $t > 0$ or not.

In order to prove the smoothing part, note that if $u_0 \in L^p_{\mathrm{loc}}(\mathbf{R})$, $p = (1-m)/2$, then

$$\int_a^x u_0^p\,dx = \int_0^y U_0^{1-p}\,dy = \int_0^y U_0^{(1-m')/2}\,dy \le C < \infty \tag{9.11}$$

for say $|y| \le 1$. We get then by Hölder

$$y \le \left(\int_0^y U_0\,dy\right)^{(m'-1)/(m'+1)} \left(\int_0^y U_0^{(1-m')/2}\,dy\right)^{2/(m'+1)},$$

so that we conclude that

$$y^{-(m'+1)/(m'-1)} \int_0^y U_0 \, dy \to \infty$$

as $y \to 0$. It has been proved in [Va84] that under this precise assumption there is no waiting time and the solution $U(y,t)$ becomes positive for all $t > 0$. It means that $u(x,t) = 1/U(y,t)$ becomes bounded for all fixed $t > 0$ near $x = a$.

The case $m = -1$ is much easier because the associated equation is the heat equation, $U_t = U_{yy}$, whose non-negative solutions are always positive for $t > 0$.

(ii) We repeat the same proof and conclude as before that $U > 0$ everywhere. But now the estimates are uniform in $x = a$.

(iii) Let us now assume that $u_0 \in L^p(\mathbf{R})$. In case $m = -1$ then $p = 1$ and we only have to check that the computation (9.11) implies that the y-interval of definition of U is bounded. Since the boundary values of U are infinity, then $U \equiv \infty$ for all $y \in \mathbf{R}$, $t > 0$. This means that $u \equiv 0$ for all $t > 0$.

In case $m < -1$, if the y-interval is bounded we are done; otherwise, we need to repeat the argument of (9.11) between x and ∞ for x large so that for very $\varepsilon > 0$ there exists $x(\varepsilon)$ such that

$$\int_{y_\varepsilon}^{\infty} U_0^{(1-m')/2} \, dy \le \varepsilon,$$

where $y_\varepsilon = y(x(\varepsilon))$. Then for $y > y_\varepsilon$ we have

$$y - y_\varepsilon \le \left(\int_{y_\varepsilon}^{y} U_0 \, dy \right)^{(m'-1)/(m'+1)} \left(\int_{y_\varepsilon}^{y} U_0^{(1-m')/2} \, dy \right)^{2/(m'+1)},$$

we conclude that

$$\lim_{y\to\infty} y^{-(m'+1)/(m'-1)} \int_0^y U_0 \, dy \to \infty.$$

Such a condition was proved by [BCP] to imply immediate blow-up for U, i.e., $U \equiv \infty$ for all $y \in \mathbf{R}$, $t > 0$. This means that $u \equiv 0$ for all $t > 0$. □

Remarks The last result is contained in [DP97]. Bertsch and Ughi [BU90] take initial data that are finite and continuous for $x \neq 0$ and infinite at $x = 0$ and prove that the solution stays unbounded at $x = 0$ for all times under the necessary and sufficient condition

$$\lim_{y\to 0} \int_{B_\rho(y)} G_\rho(x-y) u_0(x) \, dx,$$

where G_ρ is the Green function of the Laplacian in $B_\rho = \{y : |y| \le \rho\}$ with zero boundary conditions.

9.3.2 Problem in several dimensions

Theorem 9.4 *The results of Theorem 9.3 are true when $n > 1$, $m \le -1$, $p = n(1-m)/2$ and u_0 is radially symmetric.*

Proof We consider data $u_0(r)$, solve the problem to get a solution $u(r,t)$ that is easily shown to be a subsolution with respect to the solution of the one-dimensional problem with the same data. We perform the Bäcklund transform for this latter situation, We only need to check the correct computation on the initial data. We have $p = n(1-m)/2$ and then

$$y = \int_0^r u(r,t)\,dr \le \left(\int_0^r u(r,t)^p\, r^{n-1} dr\right)^{1/p} \left(\int_0^r r^{-\frac{n-1}{p-1}} dr\right)^{1-1/p}.$$

Working out the exponents and taking into account that $u_0 \in L^p_{loc}(\mathbf{R}^n)$ we get

$$y \le C(r) r^{-(1+m)/(1-m)},$$

and since $\int_0^y U_0(y)\,dy = r$ we get again

$$y^{-(m'+1)/(m'-1)} \int_0^y U_0\,dy \to \infty,$$

which allows us to conclude as before. □

In view of this result the following problem is posed.

Open problem Is the weak smoothing effect from $UL^p_{\mathrm{loc}}(\mathbf{R}^n)$ into $L^\infty(\mathbf{R}^n)$ always true on and above the critical line for $n \ge 1$ and $p > 1$?

This would immediately imply the simpler weak smoothing effect $L^p(\mathbf{R}^n)$ into $L^\infty(\mathbf{R}^n)$.

9.3.3 *The local effect when $m > -1$*

We can use a similar technique to prove the local smoothing effect from $L^1_{\mathrm{loc}}(\mathbf{R})$ into $L^\infty_{\mathrm{loc}}(\mathbf{R})$ that holds for $m > -1$. This effect is proved in [ERV88], Lemma 6.3, by a completely different technique.

Theorem 9.5 *The following results are true when $n = 1$, $-1 < m < 1$:*

(i) If $u_0 \in L^1_{\mathrm{loc}}(\mathbf{R})$, then $u(t)$ is locally bounded in $Q = \mathbf{R} \times (0,\infty)$ and more precisely,

$$u(x,t) \le C\left\{t^{1/(m+1)} M(x,R)^{2/(m-1)} + (t/R^2)^{1/(1-m)}\right\}, \qquad (9.12)$$

where $M(x,R) = \int_{B_R(x)} u_0(s)\,ds$ and $C = C(m) > 0$.

(ii) A global weak smoothing effect is true for the one-dimensional FDE in that range, going from $UL^1_{\mathrm{loc}}(\mathbf{R})$ into $L^\infty(\mathbf{R})$.

Proof (ii) is an immediate consequence of (i), so let us prove this one.

We first assume without loss of generality that $x = 0$. We may also assume that u_0 is symmetric and larger than ε by virtue of the argument done at the beginning of the proof of Theorem 9.3.

We still transform the solution according to the BT to obtain a function $U(y,t) = 1/u(x,t)$. We observe that obtaining an upper bound for $u(x,t)$ is equivalent to obtaining a lower bound for $U(y,t)$. Notice also that symmetry implies that $x = 0$ goes into $y = 0$ for all t. Therefore, we want to estimate $U(0,t)$ from below. Recall finally that distance transforms into mass and mass into distance in BT.

We will use the fact that U solves the conjugate FDE (9.10) with exponent $m' = -m \in (-1,1)$. Let us fix $R > 0$ and put $M = M(0,R)$. The initial data of U in the interval $I = (-M/2, M/2)$ are symmetric and have a mass equal to R in both directions. We have no information of what happens for $|y| \geq M/2$. Comparison of concentrations as proved in [Va04b] says that the minimum estimate is obtained when the mass of I is located as far as possible, i.e., when it is a Dirac delta of size R located at $y = M/2$, and by symmetry, the same mass at $y = -R$. But using only one of them we get the explicit lower estimate of $U(0,y)$ in terms of the ZKB solution with mass R located at distance $d = M/2$. Hence, using formula (2.21) we get

$$U(0,t)^{1-m'} \geq \frac{t}{a\,(t/M'^{1-m'})^{2/(1+m')} + k'|x'|^2},$$

where $M' = R$ and $x' = M/2$. This means that

$$u(0,t)^{1+m} \leq \frac{k\,M^2}{t} + \frac{a\,t^{(1+m)/(1-m)}}{R^{2(1+m)/(1-m)}}.$$

Raising to the power $1/(1+m)$ the estimate follows. □

Remark This technique may seem at first glance very far away from the direct comparison used in Chapters 2. to get best estimates. Note however that after performing the BT we are reduced to a worst case comparison with a ZKB solution. It is true that we do not obtain the best constant.

9.4 End-points of the critical line

Let us review what we know about the cases $m = m_c$, hence $p_* = 1$. We wonder if there is an effect from L^1 into L^∞ and of which kind.

(i) If $n \geq 3$, the answer is No, and the counterexample is constructed in the proof of Theorem 5.17.

(ii) The answer is Yes when $n = 2$, $m = m_c = 0$, as proved in [REV97]. There is a caveat, the L^1–L^∞ estimate is not uniform for all solutions of the same integral, but depends on the particular solution. This is precisely the situation that we call a *weak smoothing effect*. See Section 8.2.2.

(iii) Finally, for the same case in $n = 1$, $m_c = -1$, we have just seen that the answer is still different, we have instantaneous extinction. For locally integrable data we have locally bounded solutions.

There is another interesting aspect that we may consider, i.e., the behaviour of the solutions $u_m(x,t)$ with supercritical exponent $m > m_c$ when $m \to m_c$, keeping the initial data fixed. Let us take $u_0 \in L^1(\mathbf{R}^n)$, $u_0 \geq 0$.

(i) This limit has no problem when $n \geq 3$, since the limit solutions exist and belong to $C([0, T) : L^1(\mathbf{R}^n))$. They even have conservation of mass. More abstractly, if we consider the semigroups $S_t^{(m)} : u_0 \mapsto u_m(t)$, then $S_t^{(m)} \to S_t^{(m_c)}$ as $m \to m_c$, $m > m_c$. This continuity was proved by Bénilan and Crandall [BC81] in a more general setting.

(ii) The case $n = 2$ has been studied in [VER96]. The convergence is similar with two main differences with the result of [BC81]: the limit equation does not have uniqueness, and $S_t^{(m)}$ converges as $m \to 0$ to the semigroup of maximal solutions described in Section 8.2.

(iii) The case $n = 1$, $m_c = -1$, is more peculiar and will be studied next.

9.4.1 *Initial layer in the limit $m \to -1$, $n = 1$*

We consider here the way in which we lose the solution of the FDE when we approach the limit $m \to -1$. We take approximations with exponent $m > -1$ and fixed initial data and estimate the width of the initial layer across which the solutions disappear.

Proposition 9.6 *Let $u_0 \in L^1(\mathbf{R})$, $u_0 \geq 0$ and let u_m be the solution of the FDE with this initial data and exponent $m = -1 + \varepsilon > -1$. As $m \to -1$ the sequence*

$$v_m(x, t') = u_m(x, (1 + m)t')$$

converges to zero uniformly on $Q_M = \mathbf{R} \times (0, \tau)$ if $\tau > \tau_0 := M^2/16$, where $M = \|u_0\|_1$.

Proof If we consider the decay obtained in Section 2.2.1 by comparison with the ZKB solutions (2.21) we get the decay

$$u(x, t) \leq c(m, 1)\|u_0\|_1^{2/(m+1)} t^{-1/(m+1)} \tag{9.13}$$

with a constant given by (2.22), so that for $m \approx -1$ we have

$$c(m, 1)^{m+1} = \frac{(1 + m)}{8\pi(1 - m)} \left\{ \frac{\Gamma(1/(1 - m))}{\Gamma(1/(1 - m) + 1/2))} \right\}^2 \approx \frac{1}{16}(m + 1). \tag{9.14}$$

Let $\varepsilon = m + 1 \to 0$. We conclude that

$$u_m^{\varepsilon}(x, t) \leq \frac{M^2 \varepsilon}{16(m + 1)\, t}.$$

Putting $t' = t/(m + 1)$ the conclusion follows. □

The variable t' is a rescaling of the usual time with respect to which evolutions proceed at slower pace when $m \approx -1$. It is therefore called a *fast time*. This result shows that the exact width of the initial layer, which is $\tau_0 = M^2/16$ with respect to the fast time, and $(1 + m)\tau_0$ in standard time. Of course, it goes to zero measured in standard time.

Moreover, in the case of the ZKB solution we have the precise limit

$$\lim_{\varepsilon\to 0} U_{1+\varepsilon}(x,\varepsilon t';M)=\begin{cases}0 & \text{if } x\neq 0 \quad \text{or} \quad t'>\tau_0\\ \infty & \text{if } x=0 \quad \text{and} \quad t'<\tau_0,\end{cases} \tag{9.15}$$

cf. formula (2.21). A bit more of work shows that

$$\lim_{\varepsilon\to 0}\int_R^{\infty} U_{1+\varepsilon}(x,\varepsilon t';M)=\min\left\{2\sqrt{t'},\frac{M}{2}\right\}.$$

Therefore, for fixed $t>0$ we conclude that

$$U_m(x,(1+m)t';M)\to (M-4\sqrt{t'})_+\delta(x)$$

in the sense of measures in **R**. For general solutions, we can prove the following mass decay estimate

Theorem 9.7 *For every $u_0\in L^1(\mathbf{R})$, $u_0\geq 0$ we have*

$$\lim_{R\to\infty}\lim_{m\to -1}\int_{-R}^{R} u_m(x,(1+m)t')dx=(m-4\sqrt{t'})_+. \tag{9.16}$$

The order in this limit is important. If it is reversed, the result is just M, a consequence of the conservation of mass for $m>-1$. Another illuminating result is the following

Theorem 9.8 *Let u_m be a family of solutions as above for $m>-1$ with the same bounded initial data of mass M, let $\varepsilon=m+1\to 0$ and let*

$$w_\varepsilon(x,t')=u_m(x,\varepsilon t')^{\varepsilon}. \tag{9.17}$$

Then

$$w_\varepsilon(x,t')\to\begin{cases}1 & \text{if} \quad 0<t'\leq\tau_0\\ \tau_0/t' & \text{if} \quad t'>\tau_0.\end{cases} \tag{9.18}$$

We refer for a proof of these two results to [ERV88], pages 1018–1025, where the whole subject of this subsection has been taken from.

9.5 Comments and historical notes

Section 9.2 The radical phenomenon of instantaneous extinction occurs for different data when $m\leq 0$, and has been studied in [Va92] and then by Daskalopoulos and Del Pino [DP97, DP99]. Prior work in one dimension was due to Herrero [H89] and then [ERV88].

Section 9.3 Local smoothing effects have not been studied in a systematic way. Symmetrization does not seem to be suited to treat them, and a good technique seems to be missing. Some results for the PME are proved in [BCP]. See Section 8.3 for logarithmic diffusion in $n = 2$.

The Bäcklund transform (BT) is well-known in the case $m = -1$, where the resulting equation is the heat equation $u_t = u_{xx}$, cf. [BK80, Ro79], i.e., it linearizes a very nonlinear problem; this does not happen in the rest of the cases. On the other hand, the BT applies equation $u_t = (u^{m-1}u_x)_x$ into itself when $m = 0$. A more detailed account of the BT, its properties, use, as well as proofs, is given in [Va04c, Va05] and references. When two solutions u and U are associated by the BT as above we say that they are *BT-conjugates*, and we talk about x and y as conjugate space coordinates. It is well known that BT is related in compressible continuum mechanics to the Lagrangian formulation of the motion.

We are posing the open problem whether there is a weak smoothing effect true on the critical line for $n \geq 2$. Our conjecture is that there is one. This result seems important and the author has the impression that it should not be difficult.

Section 9.4 The FDE with exponent $m = -1$ has been proposed in [Ro79] to model heat conduction in solid hydrogen.

The calculation of the initial layer is taken from [ERV88]. The reader is advised to compare this limit situation with the case $m \to 0$ in dimension $n = 2$ treated in Subsections 5.5.4 and 8.2.1.

The limit of the ZKB solutions in formula (9.15) should be compared with similar calculations in Subsection 5.5.4 for $n = 2$ and Exercise 5.1. for $n \geq 3$. Note the difference of the time scales in the three cases

10
Summary of main results for the PME/FDE

Our main contribution contained in the preceding sections is the systematic derivation of certain estimates. They provide two kinds of information:

(i) On the one hand, we obtain *smoothing effects*, which generally speaking means that the flow creates regular solutions from non-regular data as time proceeds, a typical feature of parabolic problems. In particular, we prove that the solutions are bounded, which is the first step in the proof of regularity properties. We derive sharp a priori estimates useful in building the qualitative theory of these evolution problems.

(ii) On the other hand, we obtain estimates of the *decay rate* of the norm of the solution in time, that explains in a quantitative way the speed at which the process starting with a certain finite norm and taking place in the whole space stabilizes to zero as $t \to \infty$. We also discuss how realistic these rates are.

A rather complete picture appears, thus showing the power of the methods and the better understanding of the equation obtained in the last decades. As a complement, we discuss in less detail the question of positivity.

Let us recall that many of the results we display seem to be new. We pay special attention at the variation of the results and effects with the main parameters m and n. This is interpreted in terms of the different diffusion power and propagation speed associated to the equation when m varies.

The idea of comparison with the heat equation and its Gaussian kernel is always present. We end the chapter with a complete review of the existence of the natural extension of this object, i.e., the existence of source solutions for the whole family PME/FDE with $m \neq 1$. The concept of background signal is introduced.

10.1 Supercritical range

To begin with, we have proved that whenever the initial data belong to the Lebesgue space $u_0 \in L^1(\mathbf{R}^n)$, then the solution belongs for all times $t > 0$ to all higher spaces, $u(t) \in L^q(\mathbf{R}^n)$, $1 \le q \le \infty$, and a quantitative statement is given of that dependence. This main effect is true precisely for m larger than a critical value $m_c = (n-2)/n$ (critical exponent). The estimate corresponds to a worst case represented by a special solution, the so-called *source solution* or *ZKB solution*, see Section 2, which happens to be self-similar and explicit. We establish the effect in the form of the explicit estimate

$$|u(x,t)| \le c(m,n)\|u_0\|^{\sigma}_{L^1(\mathbf{R}^n)}\, t^{-\alpha} \tag{10.1}$$

with exact exponents, and we calculate the *best constant* $c(m,n) > 0$. Such estimates are not difficult to obtain in the linear case $m = 1$, i.e., the heat equation, since the solution is obtained explicitly by convolution of the initial data with the Gaussian kernel

$$E(x,t) = \frac{M}{(4\pi t)^{n/2}} \exp\left(-\frac{x^2}{4t}\right), \tag{10.2}$$

in the form $u(x,t) = u_0(x) * E(x,t)$ (convolution affects the space variable). Such an approach is useless in the nonlinear cases, $m \neq 1$, hence the need for completely different methods in present work.

Expanding on the issue, we show how to treat the whole family of smoothings from $L^p(\mathbf{R}^n)$ into $L^q(\mathbf{R}^n)$ with $1 \le p \le q \le \infty$. We introduce the smoothings from the Marcinkiewicz spaces M^p into L^∞, which turn out to be the natural setting for our worst-case approach.

The limit $m \to 1$ allows us to recover in a new way the estimates for the heat equation.

An estimate of the form (10.1), valid for a whole class of functions and for all times, and using only the information of the initial norm is called a *strong smoothing effect*, (SSE). We discuss the relation between SSE and the scaling properties of the equation. Though the derivation of SSEs is our main goal, they do not always exist. Consequently, the concept of *weak smoothing effect* (WSE) is introduced. Interesting estimates of this text take the form of a WSE.

We also discuss the results that can be obtained by comparing different nonlinearities φ. We obtain estimates on the positivity of solutions after a certain time in an intrinsic form, i.e., depending on the value of the local mass and the time elapsed. A fundamental difference appears: while for $m \le 1$ solutions become immediately positive, these is guaranteed to happen for $m > 1$ only after a possible waiting time.

Finally, we briefly address the question of error estimates and continuity of the PME/FDE semigroup with respect to mixed norms. All this material is developed in Chapters 2–4.

10.2 Subcritical ranges

Range $0 < m < m_c$ The previous analysis is valid for the subcritical fast-diffusion case $m < m_c$, but now we find the following essential exponent restriction: $n(m-1) + 2p > 0$. We call the points (m, p) such that $n(m-1) + 2p = 0$ the *critical line*. Note that since we work with $p \ge 1$, the study of this line deals with exponents m in the region $m \le m_c$.

The subcritical range $m < m_c$ offers two interesting novelties. Thus, on the critical line we find an important qualitative phenomenon, **extinction in finite time**. We analyse it by means of concentration comparison in Marcinkiewicz spaces. The use of Marcinkiewicz spaces in smoothing and extinction results is another novelty arising in the development of the technique. The worst-case model for extinction is an explicit solution, that we call the *explicit extinction model*, given by formula (5.2). Note that in the case of extinction, the explicit solution shows that we may expect extinction in finite time $T > 0$, but not a smoothing effect in the previous sense.

There is an even more surprising novelty. When p lies below the critical line, i.e., for $1 < p < n(1-m)/2$ (so that necessarily, $n \geq 3$ and $0 < m < (n-2)/n$) we have a *backwards effect*. By this we mean that data $u_0 \in M^p(\mathbf{R}^n)$ (in particular, $L^p(\mathbf{R}^n)$) produce solutions $u(t) \in L^1(\mathbf{R}^n)$, and a precise decay rate with best constant is found for this effect.

Range $m \leq 0$ The study of the FDE can be extended to the range $m \leq 0$, on the condition of rewriting the equation in a suitable form, $u_t = \nabla \cdot (u^{m-1}\nabla u)$, see (2.1) below. The main novelty in that area is the occurrence of *instantaneous extinction* when we are below the critical line. In this case, the solution corresponding to some non-trivial data (and properly defined as a limit of a suitable construction process) becomes identically zero for all $t > 0$. Therefore, we are in the presence of a special kind of non-existence result.

We use the trick of comparison of different diffusivities when we discuss an application to the study of a critical fast-diffusion problem, $u_t = \Delta \log u$, which has an interest in Riemannian geometry [VER96]. The logarithmic diffusion equation has very important problem of non-uniqueness due to flux at infinity that we illustrate by constructing a whole family of self-similar solutions.

We complete our analysis with a closer look at the critical line and an introduction into the study of *local smoothing effects*. We establish powerful smoothing estimates of weak local type in the case of one space dimension, and also for logarithmic diffusion if $n = 2$.

10.3 Evolution of Dirac masses. Existence of source solutions with a background

As a topic that sums up nicely our interest in special solutions, we may conclude this part of the text by addressing the problem of classifying the solutions of the PME/FDE equation with initial data a Dirac mass, so-called *source solutions*. There is no problem in that direction when $m \geq m_c$ in any dimension $n \geq 1$ since the ZKB solutions provide a clear answer to the problem. In fact, these solutions also provide the asymptotic behaviour of all non-negative solutions with finite integral.

The situation needs some further investigation in the case $m \leq m_c$. First of all, the ZKB solutions do not exist. Next we prove that they cannot exist and answer the question of the evolution of the PME flow with initial data a Dirac mass by approximation and passage to the limit. We propose the approximation scheme consisting in solving the equation with data

$$u_{0,\varepsilon}(x) = M\,\rho_\varepsilon(x), \tag{10.3}$$

where $\rho_\varepsilon > 0$ is a nice smooth function. In principle we would like to take a nice convolution kernel with integral normalized to 1. Solutions $u_\varepsilon(x, t; M)$ exist for such approximate problems if $m > 0$ and in that case we want to pass to the limit

$$U(x, t; M) = \lim_{\varepsilon \to 0} u_\varepsilon(x, t; M). \tag{10.4}$$

We call any limit of such approximations a *limit solution* (LS). However, we cannot solve the approximate problems when $m \le 0$ if we take integrable initial data, because of the phenomenon of instantaneous extinction. Therefore, we introduce a second kind of approximation which has initial data of the form

$$u_{0,M,c}(x) = M\,\rho_\varepsilon(x) + c, \tag{10.5}$$

for some constant $c > 0$. This constant is called the *background constant*, and we refer to those data as approximating the point mass with a background signal. We will solve this problem to get solutions $u_{\varepsilon,c}(x, t; M)$. We then to pass to the limit $\varepsilon \to 0$ in these solutions to get a candidate solution to the PME/FDE equation with data

$$u_{0,M,c}(x) = M\,\delta(x) + c. \tag{10.6}$$

If this second programme can be performed and the limit solutions are called $U_c(x, t; M)$, we are interested in the iterated limit

$$U^*(x, t; M) = \lim_{c \to 0} U_c(x, t; M) = \lim_{c \to 0} \lim_{\varepsilon \to 0} u_{\varepsilon,c}(x, t; M). \tag{10.7}$$

We are also interested in knowing the options that are open to us when we try to perform the double limit in any admissible way in which $(\varepsilon, c) \to (0+, 0+)$. If the double limit exists we call that limit the source solution. Otherwise, $U \ne U^*$ and we say that the source solution is not well-defined. If the source solution is well-defined, we say that it is a standard source solution if it belongs to $L^1_{\text{loc}}(Q)$.

Theorem 10.1 *If $n \ge 3$ the source solution is well-defined for $m > 0$. It is standard if $m > m_c$ and a stationary Dirac mass if $0 < m \le m_c$. If $m \le 0$ it is not well-defined.*

Proof Let us make the analysis case by case.

Case $m > m_c$, $n > 1$ There is no difficulty in solving all the approximate problems in a unique way in suitable classes of weak solutions. The approximate solutions are globally defined in time. The double limit produces the ZKB.

$0 < m \le m_c$ The plain approximation problem can be solved and the solutions $u_\varepsilon(x, t; M)$ exist, but they vanish in finite time if $m < m_c$, though they exist globally if $m = m_c$. They are positive and smooth until they vanish. We then take the limit of this process. As indicated in Exercise 5.1, we can prove that the limit solution thus obtained is exactly

$$U(x, t) = M\,\delta(x) \tag{10.8}$$

for all times $0 < t < \infty$. We recall that the result is based on a direct use of the scaling method.

When we consider the approximations with a constant background signal we still have an approximate solution $u_{\varepsilon,c}(x,t) > c$ for all times and in the limit we get the following result

$$U_c(x,t) = M\,\delta(x) + c, \tag{10.9}$$

for all times $0 < t < \infty$. This follows because of the slower rate of diffusion that allows us to apply the symmetrization-and-concentration comparison result to the equations satisfied by $v = u_{\varepsilon,c} - c$ and u_ε. We conclude that the double limit is the stationary Dirac mass at the origin, which should be taken as the physical solution of the diffusion problem. Note that this introduces the concept of measure-valued solution of a diffusion problem.

Case $m \le 0$ The problem with zero background does not admit non-trivial solutions, so that the limit of the approximations gives

$$U(x,t;M) = 0 \qquad \text{for all } x \in \mathbf{R}^n,\ t > 0. \tag{10.10}$$

This is a strong form of the instantaneous collapse already proved for L^1 functions.

However, the approach to the problem with a non-zero background suggests otherwise. Indeed, for solutions such that $u \ge c > 0$ we may use comparison of concentrations with any of the equations with $m \in (0, m_c)$ to conclude that they are more concentrated for $m \ge 0$ (once the diffusivities are renormalized up to a constant that changes time). In the limit we get the same conclusion as there

$$U_c(x,t) = M\,\delta(x) + c, \tag{10.11}$$

which holds for all $c > 0$. The iterated limit is thus

$$U(x,t;M) = M\,\delta(x) \qquad \text{in } Q. \tag{10.12}$$

It is then clear that the double limit does not exist and we cannot talk about a unique concept of source solution in this range of FDE equations. We may ask the question, what are the intermediate situations? □

Theorem 10.2 *If $n = 2$ there exists a standard source solution for $m > 0$, the source solution is well-defined but singular for $m = 0$ and it is not well-defined for $m < 0$.*

Proof There is nothing new in the case $m > 0$. We have studied in great detail the log-diffusion case $m = 0$ without background and we know that

$$U(x,t;M) = (M - 4\pi t)_+ \delta(x) \qquad \text{in } Q. \tag{10.13}$$

The case with a background has to be equally or more concentrated, so that there must be a Dirac delta at least for the same time with equal or less decay rate. It is proved in [V05b] that the situation is precisely the same. We conclude that in this case the double limit exists and the definition of measure-valued source solution can be proposed.

Case $m < 0$ The phenomenon of instantaneous extinction for finite mass solutions implies that the limit solution without background $U(x,t;M)$ is identically zero. On the other hand, the problems with a background can be compared with the problem with $m = 0$ in terms of diffusivities and concentration. We conclude that

$$U_c(x,t) = M\,\delta(x) + c, \tag{10.14}$$

so that the double limit does not exist and we do not have a source solution. □

Theorem 10.3 *If $n = 1$ there exists a unique standard source solution for $m > 0$, non-unique standard solutions if $-1 < m \leq 0$, while the source solution is identically zero for $m \leq -1$.*

Proof There is nothing new in the case $m > 0$. The source solution exists as we know for $-1 < m \leq 0$, but uniqueness needs the restriction of working with maximal solutions in this range. If we eliminate such a restriction we land (always for background $c = 0$) on a problem of non-uniqueness due to the possibility of imposing flux data at $x = \pm\infty$, and this can produce in the limit source solutions which extinguish in finite time. Such non-uniqueness does not affect the solutions of the problems with positive background, so that the iterated limit (10.7) is always the maximal source solution, whose explicit formula we have given.

Case $m \leq -1$ This is a case that can be analysed by means of the Bäcklund transform (BT) that shows that there is extinction in zero time for finite mass solutions, hence the limit solution without background $U(x,t;M)$ is identically zero. On the other hand, the limits with background correspond to conjugate solutions of the heat equation (if $m = 1$) or of the PME with exponent $m' = -m > 1$ (if $m < -1$). The details are easy after the work with the BT in Subsection 9.3.1 so we give only a sketch of the proof and ask the reader to fill in the details as an exercise:

(i) Take initial data $u_{0k}(x) = c + M\varphi_k(x)$ where $c > 0$ and $\varphi_k(x)$ is a sequence of approximations of the delta by smooth and symmetric functions of shrinking compact support. Show that there is a solution of the FDE with $m \leq -1$ and that there is a well-defined BT which produces solutions $U_k(y,t)$ of the PME with $m' \geq 1$.

(ii) Show that as $k \to \infty$ these solutions converge to the solution with initial data

$$U_0(y,t) = 0 \quad \text{for } |y| < M/2, \quad = 1/c \text{ otherwise.} \tag{10.15}$$

(iii) Check that for $m = -1$, $m' = 1$ such a solution is positive and finite for all $t > 0$. Hence, the original solution is finite for all $t > 0$ and there is no singularity (there is a smoothing effect). Conclude that the source solutions with background exist for $m = -1$ and they are smooth and bounded for all $t > 0$.

(iv) Use the property of finite propagation of the PME to show that for $m' > 1$ the solution of the PME with data (10.15) is composed for small times of two bodies separated by an empty region in the middle. The free boundaries have the form

$$s_l(t) = -\frac{M}{2} + a(m)c^{-(m'-1)/2}t^{1/2}, \quad s_r(t) = \frac{M}{2} - a(m)c^{(-m'-1)/2}t^{1/2}. \tag{10.16}$$

They meet at $y = 0$ at a time $T = 4a^2M^2c^{-(1+m)}$. Conclude by inverting the BT that the source solutions with background exist for $m < -1$ and have a singularity that stays in time for $0 < t < T$ that loses mass continuously, since the mass at zero is given by

$$M(t) = M - 2a(m)c^{(1+m)/2}t^{1/2}. \tag{10.17}$$

For $t > T$ it becomes a smooth bounded function, a further case of delayed regularity.

(v) Show that when $c \to 0$ the solutions of step (ii) go to infinity everywhere for all $m \geq -1$, so that u goes to zero. Conclude that the source solution is identically zero. □

10.4 Comments and historical notes

Smoothing effects were made popular by the works of Bénilan, from whom the author learnt the basic ideas. Another influential work of the time was Véron's [Ve79]. Extensive information on asymptotic regimes in fast diffusion is contained in [Ki93], rich in results and ideas though the derivations are formal.

Our results cover all possibilities of forward or backward smoothing effect. We propose the labels strong smoothing effect and weak smoothing effect, which seem to be new. Though the actual difference has been known by experts in various contexts, it had not been clarified to our knowledge.

Section 10.3 This section may be used by the reader to review topics of previous chapters. The idea of a background temperature is quite natural in heat propagation problems. The idea has been made famous by the discovery of the background temperature of the Universe by Penzias and Wilson in 1965.[1] Most of the material is new and the section ends with a second example of delayed regularity.

Other problems We have skipped consideration of the many results that are known for the evolution of PME/FDE flows in bounded domains with different boundary conditions. Thus, the problem with zero Dirichlet conditions is described in the survey

[1] The cosmic microwave background radiation (CMB) is a form of electromagnetic radiation that fills the whole of the Universe. It has the characteristics of black body radiation at 2.725 K.

paper [Va04], the problem with Neumann data is studied by Alikakos and Rostamian in [AR81], while the evolution on Riemannian manifolds is dealt with by Bonforte and Grillo in [BG05]. These topics are covered in the monograph [V06] which contains results and references to related problems. We have also skipped the behaviour of solutions with initial data not in L^p or M^p spaces (e.g., growing data). Another forgotten subject is the special properties of solutions with changing sign, a less studied topic for which there are partial results, see e.g., [BHi91] or [KV91]. All of these problems have traits in common with the present work but on the other hand contain very peculiar features.

Part III

Extensions and appendices

We devote a chapter to apply the foregoing ideas to the p-Laplacian evolution equation and the doubly nonlinear evolution equation.

We collect in the appendices an amount of additional information on the content of the book. We end with an appendix of comments and extensions.

11

Evolution equations of the p-Laplacian type

The combination of symmetrization, mass comparison, and time discretization can be applied to obtain useful estimates in the theory of a rather wide family of problems, more or less related in their mathematical properties to the porous medium type discussed above. Already in 1982, we showed in [Va82b] that the same symmetrization approach applies to the evolution p-Laplacian equation (PLE for short),

$$u_t - \operatorname{div}(|Du|^p Du) = f.$$

The use of the p-Laplacian operator in the diffusion term arises in several applications where the standard parabolic operator is replaced by a nonlinear diffusion with so-called gradient-dependent diffusivity. Such models appear in non-Newtonian fluids, turbulent flows in porous media, glaciology and other contexts.

Further extension to the so-called doubly nonlinear diffusion equation (DNDE),

$$u_t = \operatorname{div}(u^\sigma |Du|^{p-2} Du),$$

is not difficult. Though a complete development along the lines of what has been done for the PME/FDE is out of question for reasons of space, the calculation of best constants for the L^1–L^∞ smoothing effect for the DNLE will be performed below in full detail.

11.1 The evolution p-Laplacian equation

Let us consider the p-Laplacian equation,

$$u_t - \operatorname{div}(|Du|^{p-2} Du) = 0, \tag{11.1}$$

without forcing term. The nonlinear second-order differential operator is usually abridged as $\Delta_p u$, so we write the equation as $u_t = \Delta_p u$. The exponent p is positive, the case $p = 2$ is the heat equation. The scale invariance of the equation is a very important property that allows us to make a study along similar lines to the porous medium equation. We have the same type of worst-case strategy, since worst case models exist, namely the Barenblatt source-type solution and the explicit extinction model solution.

We will review the main facts that we need to develop a theory similar to what we have done in previous sections for the PME. A more detailed account of these preliminaries can be found in Section 8 of [Va04b]. It is a well-known fact that a p-Laplacian operator can be suitably defined so that it is m-accretive in $L^1(\mathbf{R}^n)$ and it has a dense domain. Basic definitions can be checked in [Va04b]. Therefore, it generates a semigroup of contractions in $L^1(\mathbf{R}^n)$. One main difference with the analysis of the PME is the fact that now the semigroup is a contraction in all $L^p(\Omega)$ spaces, $1 \le p \le \infty$. Consequently, the error estimates are much better than the ones of Chapter 4 for the PME.

Many of the considerations apply to the equations with gradient-dependent diffusivity of more general form, like

$$u_t - \operatorname{div}(a(|Du|)Du) = 0, \tag{11.2}$$

where a is a non-decreasing real function with suitable growth assumptions. They are usually called *gradient-dependent diffusions.* Equation (11.1) is the most representative model in that class. We will refrain from discussing the general setting for reasons of space.

11.2 The doubly nonlinear diffusion equation

Indeed, we may combine the nonlinearities of the PME and the PLE and consider the so-called doubly nonlinear diffusion equation (DNLE) without much extra formal effort. Its standard form is

$$u_t - \operatorname{div}(|D(u^m)|^{p-2}D(u^m)) = 0, \tag{11.3}$$

with $m > 0$, $p > 1$, or equivalently

$$u_t - \operatorname{div}(u^\sigma |Du|^{p-2}Du) = 0, \tag{11.4}$$

where $\sigma = (m-1)(p-1)$ and we have absorbed a factor m^{p-1} into the time variable. We will always use this latter form.

The construction of the evolution semigroup offers no special novelties. The semigroup is now a contraction in $L^1(\Omega)$. The semigroup is still bounded in $L^p(\mathbf{R}^n)$ for $p > 1$. In particular, corresponding to data $u_0 \in L^1(\mathbf{R}^n)$ we can obtain a unique mild solution $u \in C([0,\infty) : L^1(\mathbf{R}^n))$, and this is the concept of solution that we will use in this brief sketch of the properties of the solutions.

We can even combine the two types of nonlinearity and consider solving the equation

$$u_t - \operatorname{div}(|D\varphi(u)|^p D\varphi(u)) = f \tag{11.5}$$

along the same lines, see the details in [Va04b].

11.3 Symmetrization and mass comparison

The symmetrization considerations we have made in Section 1.3 for the filtration equation can be extended to the present setting. We then have a result similar to Theorem 1.3.

Theorem 11.1 *Let u be the mild solution of equation* (11.5) *posed in $Q_T = \mathbf{R}^n \times (0, T)$ with data $u_0 \in L^1(\mathbf{R}^n)$, and $f \in L^1(Q_T)$. Let v be the solution of a similar problem with radially symmetric data $v_0(r) \geq 0$, and second member $g(r, t) \geq 0$. Assume moreover that*

$$(i) \quad u_0^* \prec v_0, \qquad (ii) \quad f^*(\cdot, t) \prec g(\cdot, t) \quad \text{for every } t \geq 0.$$

Then, for every $t \geq 0$ we have

$$u^*(\cdot, t) \prec v(\cdot, t). \tag{11.6}$$

In particular, for every $p \in [1, \infty]$ we have comparison of L^p norms,

$$\|u(\cdot, t)\|_p \leq \|v(\cdot, t)\|_p\,. \tag{11.7}$$

11.4 Source-type solutions

For ease of reference, we treat separately the PLE and the DNLE.

11.4.1 p-Laplacian source solution

To begin with, the source solution, or Barenblatt solution, is explicitly given for $p > 2$ by

$$U(x, t; M) = t^{-\alpha} F(x/t^{\alpha/n}), \quad F(\xi) = \left(C - k\,|\xi|^{\frac{p}{p-1}}\right)_+^{\frac{p-1}{p-2}} \tag{11.8}$$

where

$$\alpha = \frac{n}{n(p-2)+p}, \quad k = \frac{p-2}{p}\left(\frac{\alpha}{n}\right)^{\frac{1}{p-1}}. \tag{11.9}$$

As in the porous medium case, it has a Dirac delta as initial trace

$$\lim_{t\to 0} u(x, t) = M\,\delta_0(x).$$

The remaining parameter $C > 0$ of formula (11.8) is free and can be uniquely determined in terms of the mass, $\int U dx = M$, which gives the following relation between M and C:

$$M = d\,C^\gamma, \quad d = n\,\omega_n \int_0^\infty (1 - k\,y^{p/(p-1)})_+^{(p-1)/(p-2)}\, y^{n-1}\,dy, \quad \gamma = \frac{(p-1)n}{p(p-2)\alpha} \tag{11.10}$$

(k and γ are functions depending only on p and n).

The source solution is still valid with simple sign changes for $1 < p < 2$ as long as α remains positive, i.e., for $p > p_c = 2n/(n+1)$. Note that while the solution has compact support in space for $p > 2$, it is positive everywhere for $p_c < p < 1$ with formula

$$U(x,t;M) = t^{-\alpha} F(x/t^{\alpha/n}), \quad F(\xi) = \left(C + |k|\,|\xi|^{\frac{p}{p-1}}\right)_+^{-\frac{p-1}{2-p}}. \tag{11.11}$$

Remark When comparing the PME with the PLE the reader should note that the diffusion term in the equation involves the m-th power of u in the former case, the $(p-1)$-th power of u in the second, while the 'number of derivatives' in x is two in the former, p in the latter case. This will help in comparing the exponents in the decay formulas. Since they come from algebraic considerations on scaling (dimensional analysis), they must respect such analogy.

11.4.2 The doubly nonlinear equation

A Barenblatt solution exists for the DNLE in the 'good range'

$$m(p-1) + (p/n) > 1 \tag{11.12}$$

that includes of course $m \geq 1$, $p \geq 2$. When, moreover, $m(p-1) > 1$, the source solution is precisely given by

$$U(x,t;M) = t^{-\alpha} F(x/t^{\alpha/n}), \quad F(\xi) = \left(C - k\,|\xi|^{\frac{p}{p-1}}\right)_+^{\frac{p-1}{m(p-1)-1}} \tag{11.13}$$

where

$$\alpha = \frac{1}{m(p-1) - 1 + (p/n)}, \qquad k = \frac{m(p-1)-1}{p}\left(\frac{\alpha}{n}\right)^{\frac{1}{p-1}}. \tag{11.14}$$

For $m(p-1) < 1$ and $m(p-1) + (p/n) > 1$ we get the same self-similar form but now with profile

$$F(\xi) = \left(C + |k|\,|\xi|^{\frac{p}{p-1}}\right)^{-\frac{p-1}{1-m(p-1)}}. \tag{11.15}$$

Finally, for $m(p-1) = 1$ the differential operator in the right-hand side is homogeneous of degree one, though the equation is still nonlinear if $m \neq 1$. We have exponential profiles that can be obtained as a limit of the preceding ones

$$F(\xi) = C \exp\left(-k\xi^{p/(p-1)}\right), \tag{11.16}$$

and now $k = (p-1)p^{-p/(p-1)}$. Note that in this case the value of the similarity exponent is $\alpha = n/p$, and the space expansion rate is $\beta = \alpha/n = 1/p$.

Remark Same observation about the numerology of the exponents as in the PLE.

11.5 Smoothing estimates, best constants and decay rates for PLE and DNLE

We outline the main points of a theory that parallels the one developed for the PME/FDE case. We concentrate more on supercritical exponents, even if the work on the subcritical range is most interesting. The results and the outline of the proofs are similar to the case of the PME. We can also continue along similar lines the study of the other topics of Chapters 2–5: general smoothing effects, extinction, and backward effects.

Three considerations are relevant at this point. First, that the analysis offers many similarities at the formal level; second, that there are a number of differences in detail, both in theory and in the values of the exponents that are worth examining and listing, since we are dealing with basic properties of quite frequent mathematical models for nonlinear diffusion; third, the same remarks apply to the p-Laplacian model and to the more general model called the doubly nonlinear diffusion equation. Therefore, at the risk of asking the reader for extra effort, we continue the analysis with the more general model.

11.5.1 The L^1–L^∞ effect

Using the existence and properties of the source solutions and the comparison theorems in the spirit of Section 2.2, we get the following result.

Theorem 11.2 *Let u be the mild solution of equation* (11.4) *in the range $m(p-1)+(p/n)>1$ with initial datum $u_0 \in L^1(\mathbf{R}^n)$. Then, for every $t>0$ we have $u(t)\in L^\infty(\mathbf{R}^n)$ and moreover there is a constant $c(m,p,n)>0$ such that*

$$0 \le u(x,t) \le c(m,p,n) M^{p\alpha/n} t^{-\alpha} \tag{11.17}$$

is valid for all $x\in\mathbf{R}^n$ and $t>0$, with $M=\|u_0\|_{L^1(\mathbf{R}^n)}$. The best constants are

$$c(m,p,n) = \left(\frac{p-1}{\lambda p}\right)^{\alpha(p-1)} \left(\frac{\alpha}{n}\right)^{\alpha} \left(\frac{n}{2(p-1)\pi^{n/2}\alpha}\,\frac{\Gamma(n/2)\,\Gamma(\gamma)}{\Gamma(n(p-1)/p)\,\Gamma(\lambda)}\right)^{\frac{\alpha p}{n}} \tag{11.18}$$

if $m(p-1)>1$,

$$c(m,p,n) = \left(\frac{p-1}{|\lambda| p}\right)^{\alpha(p-1)} \left(\frac{\alpha}{n}\right)^{\alpha} \left(\frac{p}{2(p-1)\pi^{n/2}}\,\frac{\Gamma(n/2)\,\Gamma(|\lambda|)}{\Gamma(n(p-1)/p)\,\Gamma(|\gamma|)}\right)^{\frac{\alpha p}{n}} \tag{11.19}$$

if $m(p-1)<1$, and

$$c(m,p,n) = \frac{(p-1)^{\mu-1}}{2p^{n-1}\pi^{n/2}}\,\frac{\Gamma(n/2)}{\Gamma(n(p-1)/p)} \tag{11.20}$$

if $m(p-1)=1$. Here, $\lambda=(p-1)/(m(p-1)-1)$ and $\mu=(p-1)n/p$.

Taking the limit as $m\to 1$, $p\to 2$ in all three expressions for the best constant, we get the well-known constant for the L^1–L^∞ effect for the heat equation, $c(1,2,n)=(4\pi)^{-n/2}$, as expected. Since the calculation is a bit lengthy and has no real novelties, we have preferred to postpone the details to Subsection 11.5.6.

11.5.2 Smoothing effects for the PLE

The application of the results to the case $m = 1$, $p \neq 2$ gives

Theorem 11.3 *Let u be the solution of equation (11.1) in the range $p > p_c$ with initial datum $u_0 \in L^1(\mathbf{R}^n)$. Then, for every $t > 0$ we have $u(t) \in L^\infty(\mathbf{R}^n)$ and moreover there is a constant $c(p,n) > 0$ such that*

$$|u(x,t)| \leq c(p,n)\, \|u_0\|_1^\sigma\, t^{-\alpha}, \tag{11.21}$$

with α given in (11.9) and $\sigma = p\,\alpha/n$. The optimal constant is attained by the Barenblatt solution.

11.5.3 General smoothing effects

The effects of M^r into L^∞ are derived from the existence of bounded solutions with initial data $U_q(x) = A\,|x|^{-n/r}$ on the condition that $pr + nm(p-1) > n$. The following holds

Theorem 11.4 *Let $p, r > 1$ be such that $n(p-2) + rp > 0$. For every $u_0 \in M^r(\mathbf{R}^n)$, $1 < r < \infty$, and every $t > 0$ we have $u(t) \in L^\infty(\mathbf{R}^n)$ and*

$$|u(x,t)| \leq c(p,n,r,\infty)\, \|u_0\|_{M^r}^{\sigma_r}\, t^{-\alpha_r} \tag{11.22}$$

with

$$\alpha_r = \frac{n}{rp + n(p-2)}, \qquad \sigma_r = \frac{rp}{rp + n(p-2)}. \tag{11.23}$$

Moreover, when $r = 1$ the space is replaced by $\mathcal{M}(\mathbf{R}^n)$. The best constants are attained by the Barenblatt solution if $r = 1$, by the solution with data $u_0(x) = A\,|x|^{-n/r}$ if $r > 1$.

We leave to the reader the proof of this result, which has no essential novelties over what has been done for the PME.

11.5.4 Extinction

When we enter the 'lower range' $m(p-1) + (p/n) < 1$, smoothing, extinction, and backward effects can be considered. The extinction situation is similar to the cases already studied: it is also tied to an explicit solution defined for $m(p-1) + (p/n) < 1$ by

$$U(x,t) = c_{m,p} \left(\frac{T-t}{|x|^p} \right)^\mu, \qquad \mu = \frac{1}{1 - m(p-1)}, \tag{11.24}$$

with

$$c(m,p,n) = \Big((p\mu)^{p-1} n/|\alpha| \Big)^\mu. \tag{11.25}$$

This function belongs to a space $M^{r_*}(\mathbf{R}^n)$ with $r_* > 1$ if we are in the aforementioned lower range, and

$$r_* = \frac{n(1 - m(p-1))}{p}. \tag{11.26}$$

The *extinction surface* for the DNLE posed in the whole space is defined by the expression

$$pr + nm(p-1) = n. \tag{11.27}$$

with parameters m, p, r.

In the p-Laplacian case, $m = 1$ and p lives in the range $1 < p < p_c = 2n/(n+1)$. This is the range in which the uniqueness theory needs for a delicate definition of the concept of solution, cf. [B6-95]. The separate variable solution has $\mu = 1/(2-p)$ and $c = c_p = |k|^{p/(2-p)}$, k given in (11.9). The extinction line in the p-Laplacian problem is $p(r+n) = 2n$.

11.5.5 Backwards effect

There is also a backwards effect from $M^r(\mathbf{R}^n)$ into $L^q(\mathbf{R}^n)$ with $1 \le q \le p$, if $m(p-1) + (p/n) < 1$ and $1 < r < r_*$, where r_* is the extinction exponent. It all depends on the study of the phase-plane analysis for the autonomous system of the DNLE (or the PLE), work that can be done with only some effort. Figure 11.1 is to be compared with Figure II.1, page 74, valid in the PME / FDE case.

We refrain from further details of different qualitative effects by lack of space, and finish this section with a practical aspect.

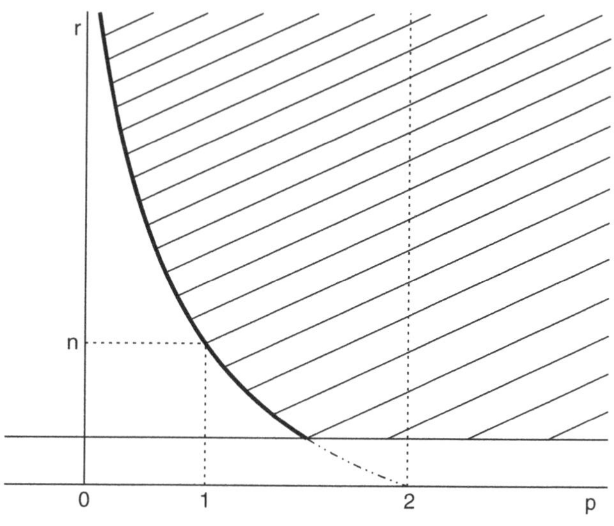

Figure 11.1 The (m, p) diagram for the PLE in dimensions $n \ge 3$. The boldface line is the critical line.

11.5.6 Calculation of the best constants for the doubly nonlinear equation

The outline follows the PME case of Section 2. The worst case with respect to the symmetrization-and-concentration comparison in the class of solutions with the same initial mass M is just the ZKB solution U with initial data a Dirac mass, $u_0(x) = M\delta(x)$. We are thus reduced to performing the computation of the best constant for the ZKB solution. When $m(p-1) \neq 1$, using formula (11.13) gives

$$\|U(t)\|_\infty = C^\lambda t^{-\alpha}, \quad \lambda = \frac{p-1}{m(p-1)-1}.$$

We will be using two more constants,

$$\gamma = \frac{n(p-1)}{p} + \lambda = \frac{n\lambda}{\alpha p}, \qquad \mu = \frac{(p-1)n}{p}.$$

While μ is always positive, λ has the sign of $m(p-1)-1$; and γ behaves likewise if $\alpha > 0$, as is presently assumed. The computation of the mass, $M = \int U(x,t)\,dx$, is

$$M = n\,\omega_n \int_0^\infty (C - k\,y^{\frac{p}{p-1}})_+^\lambda\, y^{n-1}\,dy, \quad M = n\,\omega_n \int_0^\infty (C + |k|\,y^{\frac{p}{p-1}})_+^\lambda\, y^{n-1}\,dy,$$

with $\pm$ according to the sign of $m(p-1)-1$, i.e., of λ. Then, $M = d\,C^\gamma\,|k|^{-\mu}$, with

$$d = n\,\omega_n \int_0^\infty (1 \pm y^{p/(p-1)})_+^\lambda\, y^{n-1}\,dy,$$

with $\pm$ according to the sign of λ as before. Noting that $\lambda/\gamma = \alpha p/n$ and $\lambda\mu/\gamma = \alpha(p-1)$, we have

$$\|U(t)\|_\infty = d^{-p\alpha/n} k^{\alpha(p-1)} M^{p\alpha/n} t^{-\alpha}.$$

The best constant is therefore

$$c(m.n,p) = d^{-p\alpha/n} k^{\alpha(p-1)}. \tag{11.28}$$

In order to obtain a more explicit expression, we have to do an exercise involving Euler's beta and gamma functions.

(i) For $m(p-1) > 1$ we obtain

$$d = n\omega_n \int_0^1 (1 - s^{p/(p-1)})^\lambda s^{n-1} ds = \frac{n\omega_n(p-1)}{p} B\left(\frac{n(p-1)}{p}, \lambda + 1\right).$$

Since $\gamma = \lambda + n(p-1)/p = \lambda n/p\alpha$, we get

$$d = \frac{2(p-1)\pi^{n/2}}{p\Gamma(n/2)} \frac{\Gamma(n(p-1)/p)\,\Gamma(\lambda+1)}{\Gamma(\frac{n(p-1)}{p} + \lambda + 1)} = \frac{2(p-1)\pi^{n/2}\alpha}{n} \frac{\Gamma(n(p-1)/p)\,\Gamma(\lambda)}{\Gamma(n/2)\,\Gamma(\gamma)}.$$

Using the value

$$k = \frac{m(p-1)-1}{p}\left(\frac{\alpha}{n}\right)^{\frac{1}{p-1}} = \frac{p-1}{\lambda p}\left(\frac{\alpha}{n}\right)^{\frac{1}{p-1}},$$

we conclude that inequality (11.17) holds with the precise constant

$$c(m,p,n) = \left(\frac{p-1}{\lambda p}\right)^{\alpha(p-1)}\left(\frac{\alpha}{n}\right)^{\alpha}\left(\frac{n}{2(p-1)\pi^{n/2}\alpha}\,\frac{\Gamma(n/2)\,\Gamma(\gamma)}{\Gamma(n(p-1)/p)\,\Gamma(\lambda)}\right)^{\frac{\alpha p}{n}}, \tag{11.29}$$

which is to be compared with formula (2.13) for $p = 2, m > m_c$.

(ii) For $m(p-1) < 1$ we have $\lambda, \gamma, k < 0$ and

$$d = n\omega_n \int_0^\infty (1 + s^{p/(p-1)})^{-|\lambda|} s^{n-1} ds = n\omega_n \frac{p-1}{p} B\left(\frac{n(p-1)}{p}, \frac{n|\lambda|}{p\alpha}\right),$$

$$d = \frac{2(p-1)\pi^{n/2}}{p\Gamma(n/2)}\,\frac{\Gamma(n(p-1)/p)\,\Gamma(|\gamma|)}{\Gamma(|\lambda|)}.$$

Therefore,

$$c(m,p,n) = \left(\frac{p-1}{|\lambda| p}\right)^{\alpha(p-1)}\left(\frac{\alpha}{n}\right)^{\alpha}\left(\frac{p}{2(p-1)\pi^{n/2}}\,\frac{\Gamma(n/2)\,\Gamma(|\lambda|)}{\Gamma(n(p-1)/p)\,\Gamma(|\gamma|)}\right)^{\frac{\alpha p}{n}},$$

which is to be compared with formula (2.14), for $p = 2, m_c < m < 1$.

(iii) The case $m(p-1) = 1$ is a bit different; it is equivalent to working in the limit $\lambda \to \infty$ of the above cases. To calculate the best constant we use formula (11.16), and the similarity exponent is $\alpha = n/p$. The computations give

$$M = Cn\,\omega_n \int_0^\infty \exp\left(-k\,y^{\frac{p}{p-1}}\right) y^{n-1}\,dy = Ck^{-\mu} n\,\omega_n \int_0^\infty \exp\left(-s^{\frac{p}{p-1}}\right) s^{n-1}\,ds$$

$$= \frac{p-1}{p} n\,\omega_n C k^{-\mu} \int_0^\infty e^{-t} t^{\frac{n(p-1)}{p}-1}\,dt = \frac{p-1}{p} n\,\omega_n C k^{-\mu}\Gamma\left(\frac{n(p-1)}{p}\right).$$

Using the value of k in this case and $n\omega_n = 2\pi^{n/2}/\Gamma(n/2)$, we conclude that $c(m,n) = C/M$ equals

$$c(m,p,n) = \frac{pk^{\mu}}{2(p-1)\pi^{n/2}}\,\frac{\Gamma\left(\frac{n}{2}\right)}{\Gamma\left(\frac{n(p-1)}{p}\right)} = \frac{(p-1)^{\mu-1}}{2p^{n-1}\pi^{n/2}}\,\frac{\Gamma\left(\frac{n}{2}\right)}{\Gamma\left(\frac{n(p-1)}{p}\right)}. \tag{11.30}$$

11.6 Comments and historical notes

We have already pointed out the similarity between the PME and the PLE in a number of works, like [Va90]. A general reference for the theory of p-Laplacian equations is DiBenedetto's book [DiB93]. See also [DF85, DH89, DH90, Zh95] among many references to the existence, uniqueness, and regularity literature on the subject.

The evolution p-Laplacian equation is one of most widely researched equations in the class of nonlinear degenerate parabolic equations, already studied by Raviart [Rav70] (along with the doubly nonlinear parabolic equation). The particular feature of equation (11.1) is its gradient-dependent diffusivity. Such equations, and their stationary counterparts, appear in different models in non-Newtonian fluids, cf. [La69], in glaciology [Ht97], turbulent flows in porous media, certain diffusion or heat transfer processes, and recently in image processing. See, in that last respect, [PM90, CL71, BV04]. The limit $p \to \infty$ is proposed in [EFG97] as providing a simplistic model for the 'collapse of an initially unstable sandpile'. The value $p = 1$ is studied in connection with image restoration in [ACM04]. The elliptic version is widely used in the calculus of variations, in connection with nonlinear elasticity and quasi-conformal mappings.

The doubly nonlinear equation has been investigated by different authors after the pioneering work of Kalashnikov, see the survey [Ka87]. It had been proposed by Leibenzon in modelling turbulent fluid problems [Le45], see also [EV86]. The application to glaciology appears for instance in [C5-02]. Concerning the theory and basic estimates there are many results by a large number of authors, among them [EV88] where we gave the first proof of asymptotic behaviour (in one space dimension).

Section 11.4 The source solution for the PLE is due to Barenblatt [Ba52b]. A very detailed study of self-similarity for the p-Laplacian equation in the fast diffusion case has been recently done by [BiV05].

Section 11.5 The derivation of the smoothing effect L^1–L^∞ with best constant by the present method was done in [Va82b], but the constants were not calculated. The smoothing results of Theorem 11.4 are derived in [HV81] and [Ve79] with data in L^r, using iteration techniques. A more precise asymptotic convergence theorem for L^1 data was first proved in [KV88].

The extinction result for the PLE was proved in [HV81] for data in L^p. In that case, the solution belongs to L^∞ for all $0 < t < \infty$, cf. [HV81], page 117. This is a weak smoothing effect. It implies that the critical case proves all the rest of the smoothing effects into L^∞ when $1 < p < 2n/(n+1)$. Finite time extinction for the fast p-Laplacian equation was studied by [QW99].

Exercises

Exercise 11.1. Prove the delayed regularity result of Chapter 6 in the case of the PLE. The proof is much easier due to the invariance of the equation under vertical displacements, cf. [HV81].

A harder problem is proving the result for the DNLE in the appropriate range.

Exercise 11.2. Take the PLE in dimension $n > 1$ with $p = n$ and write the formula for the Barenblatt solution. Now take the limit $m \to 0$ to find another example of the curious phenomenon of Dirac masses decaying without diffusing.

Exercise 11.3. In order to compare the critical value p_c of the PLE with m_c of the PME we may write

$$m_c = 1 - \frac{2}{n}, \qquad p_c = 2 - \frac{2}{n+1}.$$

We ask the reader to explain the difference.

Appendices

Appendix I. Some analysis topics

AI.1 Some integrals and constants

We list some of the integrals that enter the calculation of the best constants in the smoothing effect.

(i) EULER'S GAMMA FUNCTION is defined as

$$\Gamma(p) = \int_0^\infty t^{p-1} e^{-t}\, dt, \qquad p > 0.$$

We have $\Gamma(p) = (p-1)\,\Gamma(p-1)$, and $\Gamma(1) = 1$, $\Gamma(1/2) = \sqrt{\pi}$. As $p \to \infty$ we have

$$\Gamma(p) \sim (p/e)^p (2\pi p)^{1/2}.$$

(ii) EULER'S BETA FUNCTION is defined for $p, q > 0$ as

$$B(p,q) = \int_0^1 t^{p-1}(1-t)^{q-1}\, dt = 2\int_0^1 s^{2p-1}(1-s^2)^{q-1}\, ds.$$

We have $B(p,q) = B(q,p)$ and the basic relation

$$B(p,q) = \frac{\Gamma(p)\Gamma(q)}{\Gamma(p+q)},$$

as well as the equivalent expressions with parameter $r > 0$

$$B(p,q) = r\int_0^1 s^{rp-1}(1-s^r)^{q-1}\, ds = r\int_0^\infty \frac{x^{rq-1}}{(1+x^r)^{p+q}}\, dx.$$

These expressions are usually found for the value $r = 2$.

(iii) VOLUME OF THE UNIT SPHERE: It is well known that

$$\omega_n = \frac{2\pi^{n/2}}{n\Gamma(n/2)} = \frac{\pi^{n/2}}{\Gamma(n/2+1)}.$$

This formula easily follows from the properties of Euler's integrals by induction.

AI.2 More on Marcinkiewicz spaces. Lorentz spaces

From the definition of Marcinkiewicz norm in Section 1.1, formula (1.13), we immediately obtain an estimate for the distribution function of f of the form:

$$\mu_f(k) \le \left(\frac{\|f\|_{M^p}}{k}\right)^p .$$

A norm in $M^p(\mathbf{R}^n)$ is obtained by taking the infimum of the constants in this estimate:

$$|||f|||_{M^p(\mathbf{R}^n)} = \inf\{C : \ \mu_f(k) \le (C/k)^p \ \forall k > 0\}. \tag{AI.1}$$

It is easily proved that the new norm is equivalent to the old one for all $1 < p < \infty$, cf. [BBC75], page 548.

In this norm the explicit function $U_p(x) = A\,|x|^{-n/p}$ of Sections 3.3 and 5.4 is not only the most concentrated one in the class of rearranged functions of equal norm, it is even the *greatest element* up to rearrangement. The existence of greatest element is an additional simplification for comparison arguments.

The same definition and norms apply when the base space is a domain of $\mathbf{R}^n$, or even a measurable subset K. One of the important functional features of $M^p(\mathbf{R}^n)$ or $M^p(\Omega)$ is the fact that they are not separable. According to [BS88], every separable Banach function space is the closure of the set of bounded functions with support in a set of finite measure. This is not the case here, and in fact, when Ω is bounded, $L^p(\Omega)$ is a non-dense subset of $M^p(\Omega)$. When Ω is unbounded or $\mathbf{R}^n$, then $L^p_{\rm loc}(\Omega) \cap M^p(\Omega)$ is a non-dense subset.

In order to clarify this fact, we introduce the following functional in $M^p(\Omega)$:

$$\mathrm{N}_p(f) = \lim_{k\to\infty} \|(|f| - k)_+\|_{M^p}. \tag{AI.2}$$

Note that the limit exists since the family $(|f| - k)_+$ is non-increasing as $k \to \infty$. Actually, the definition only needs f to be such that $(|f| - k)_+ \in M^p(\mathbf{R}^n)$ for some $k > 0$.

Lemma AI.1 N_p *is a continuous seminorm in* $M^p(\Omega)$. *Given* $f \in M^p(\Omega)$ *we have* $\mathrm{N}_p(f) = 0$ *if* $f \in M^p_0(\Omega)$, *which is defined as the closure of* $C^1_c(\Omega)$ *in* $M^p(\Omega)$. *If* Ω *is bounded, the converse is true.*

Proof The fact that N_p is a seminorm and is continuous are both clear. Let us prove next that $\mathrm{N}_p(f) = 0$ for every $f \in M^p_0(\Omega)$. In the case where f is C^1_c it is bounded, and then $\mathrm{N}_p(f) = 0$. The result follows for $M^p_0(\Omega)$ by density.

On the contrary, if $N_p(f) = 0$, for every $\varepsilon > 0$ there is a value k such that $\|f_k\|_{M^p} \le \varepsilon$ with $f_k = (|f| - k)_+ \mathrm{sign}(f)$. Now, $f - f_k \in L^\infty(\Omega) \cap M^p(\Omega) \subset L^p(\Omega)$, since Ω has bounded measure. The result follows. □

Note that the limit $\mathrm{N}_p(f)$ is not zero in the case of example (1.14); actually, the limit is still $\mathrm{N}_p(U_p) = A\,\kappa_p$.

Lorentz spaces $L^{p,q}(\Omega)$, $1 \le p, q < \infty$, are defined by means of the norms

$$\|f\|_{p,q} = \left(\int_0^{|\Omega|} s^{\frac{q}{p}-1} f_*^q(s)\,ds\right)^{1/q} = n^{\frac{1}{q}} \omega_n^{\frac{1}{p}} \left(\int_0^R r^{\frac{nq}{p}-1} (f^*)^q(r)\,dr\right)^{1/q} < \infty,$$

where R is defined by $|B_R| = |\Omega|$. Here, f_* denotes the one-dimensional symmetric representation of a function $f \in \mathcal{L}_0$ defined by means of

$$f_*(s) = f^*(r), \quad s = \omega_n r^n. \tag{AI.3}$$

Then, f_* is defined in the interval $[0, |\Omega|]$, with $|\Omega| = \text{meas}\,(\Omega)$. Notice that

$$f_*(s) = \inf\{t \ge 0 : \mu(t) < s\}, \tag{AI.4}$$

which makes f_* a generalized inverse of μ_f.

It is easy to see that $L^{p,p}(\Omega)$ coincides with $L^p(\Omega)$. The $L^{p,q}(\Omega)$ form an increasing family of separable spaces as q increases. In the limit $q \to \infty$ the norm becomes

$$\|f\|_{p,\infty} := \text{ess sup}\,\{s^{1/p} f_*(s)\} < \infty.$$

Comparing this with (AI.1), we see that $L^{p,\infty}(\Omega) = M^p(\Omega)$.

AI.3 Morrey spaces

We have seen these in the study of extinction. Here is the definition.

Definition AI.1 Let $p \in (1, \infty)$ and $\lambda \in (0, n)$. The function $f \in L^1_{\text{loc}}(\Omega)$ is said to belong to the Morrey space $\widetilde{M}^{p,\lambda}(\Omega)$ if

$$\|f\|_{L^{p,\lambda}(\Omega)} = \sup_{\rho > 0, x \in \Omega} \{\rho^{-\lambda} \int_{B_\rho(x) \cap \Omega} |f|^p\,dx\} < +\infty, \tag{AI.5}$$

where $B_\rho(x)$ denotes the n-dimensional ball of radius ρ centred at the point x.

AI.4 Maximal monotone graphs

The nonlinearity in the filtration equation is in principle a C^1 monotone function, so that the equation may be written as

$$u_t = \nabla \cdot (\varphi'(u)\,Du)$$

and the formal parabolic character is easy to recognize. If $\varphi'(u) > 0$ for every u, than the problem is actually parabolic. But the applications point to the interest in dealing with extensions of the nonlinearity to include non-strictly increasing or multi-valued functions. In order to encompass all these applications the concept of *maximal monotone graph* is used, m.m.g. for short. We think of φ as a graph consisting of pairs $(u, v) \in \mathbf{R}^2$, with u in a domain $D(\varphi)$, which is an interval of the real line. The graph is monotone if for every two graph pairs (u_1, v_1) and (u_2, v_2) we have

$$(u_1 - u_2)(v_1 - v_2) \geq 0.$$

It is maximal if it cannot be extended further. A continuous monotone function defined in $\mathbf{R}$ is a good example of m.m.g. For a discontinuous monotone function the key point is that we have to add all the vertical segments corresponding to the multi-valued options, and it has to be defined in a maximal interval $D(\varphi)$. This interval is not necessarily $\mathbf{R}$. Typical maximal monotone graphs appearing in the proof of estimates for the types of PDEs we deal with in this book are the *sign* function

$$\text{sign}(s) \begin{cases} = 1 & \text{for } s > 0, \\ = -1 & \text{for } s < 0, \\ = [-1, 1] & \text{for } s = 0; \end{cases}$$

and also its positive part, denoted by $\text{sign}^+(s)$, where we modify $\text{sign}(s)$ so that $\text{sign}^+(s) = 0$ for $s < 0$ and $\text{sign}^+(0) = [0, 1]$.

One of the main advantages of this generality is the fact that the inverse of a m.m.g. is again a m.m.g.; actually, both graphs are symmetric with respect to the main bisectrix in $\mathbf{R}^2$. The standard and somewhat awkward notation when using multivalued operators is set inclusion, so that when (a, b) is a point in the graph φ we write $b \in \varphi(a)$ instead of $b = \varphi(a)$, since generally $\varphi(a)$ is not a singleton.

Maximal monotone operators in Hilbert spaces have been studied in full detail by Brezis in [Br73]. A review of the application of maximal monotone operators to the theory of the PME is contained in Chapter 10 of [V06].

Appendix II. Particles and speeds

AII.1 Lagrangian approach in diffusion

The Lagrangian approach to diffusion equations consists in assuming that a solution $u(x,t) \ge 0$ represents the density at time t of a mass distribution extending to a certain domain, here the whole space, $x \in \mathbf{R}^n$. The mass contained in a measurable set is calculated as

$$M(t;\, A) = \int_A u(x,t)\, dx,$$

so that $M(t) = \int_{\mathbf{R}^n} u(x,t)\, dx$ represents the total mass of this distribution at time $t \ge 0$. The mass conservation law of continuum mechanics is

$$\partial_t u + \nabla(u\, \mathbf{v}) = 0, \tag{AII.1}$$

where $\mathbf{v}$ is the pointwise velocity of the particles that make up the medium. Agreement of this formula with the PDE/FDE equation $u_t = \nabla \cdot (u^{m-1}\nabla u)$ leads to postulate a formula for the velocity of the form

$$\mathbf{v} = -u^{m-2}\nabla u = -\frac{1}{m-1}\nabla u^{m-1}. \tag{AII.2}$$

Note that for the HE the last formula reads $\mathbf{v} = -\log u$. This way of looking at the equation is quite typical of the modelling of flows in porous media for $m > 1$, cf. [Mu37, Ar86], but there is no problem in extending the idea to $m < 1$, the main difference being that $m - 1 < 0$, so we are taking the gradient of a negative power of the density u.

A similar reasoning applies to the p-Laplacian case. Equation $u_t = \operatorname{div}(|Du|^{p-2}Du)$, can be written in the form (AII.1) if we define

$$\mathbf{v} = -u^{-1}|Du|^{p-2}\nabla u = -|DP|^{p-2}\nabla P, \quad P = \frac{p-1}{p-2}u^{(p-2)/(p-1)}. \tag{AII.3}$$

Here, P is called the *nonlinear velocity potential*, cf. [EV86]. In the same way, the doubly nonlinear equation (11.4) has velocity given by

$$\mathbf{v} = -u^{\sigma-1}|Du|^{p-2}\nabla u = -|DP|^{p-2}\nabla P, \quad P = \frac{p-1}{p+\sigma-2}u^{(p+\sigma-2)/(p-1)}. \tag{AII.4}$$

Note that $p + \sigma - 2 = m(p-1) - 1$ in the notation of (11.3).

If we want to find the particle trajectories, we have to integrate the kinematic equation

$$\frac{dx}{dt} = \mathbf{v}(x,t), \tag{AII.5}$$

Often, we apply the particle approach to obtain the way in which particles that are initially located far away diverge to infinity with time. This uses the knowledge of the specific asymptotic behaviour of the solutions as $|x| \to \infty$; from that, information on the behaviour of $\mathbf{v}$ is derived, which is then used to estimate the trajectories.

AII.2 The ZKB solutions and similar

Let us take the source-type ZKB solutions of the PME/FDE as example. It is very easy to check that the particle velocity, given by formula (AII.2), is in this particular case described by the linear dependence

$$\mathbf{v} = \beta x/t. \tag{AII.6}$$

This simple relation between the velocity and the density is characteristic of the ZKB solutions as the reader may check by himself.[1]

We now integrate the kinematic equation for the particle trajectories (AII.5) to obtain a divergent behaviour of the form

$$x(t) = x(1)\,t^{\beta}, \tag{AII.7}$$

describing the non-Gaussian way in which particles spread (note that $1/2 < \beta < \infty$ if $1 > m > m_c$, while $1/2 > \beta$ for $m > 1$). Both the PME and the FDE equation are examples of anomalous diffusion in the sense that is used in the literature. Of course, we recover the Brownian motion exponent $\beta = 1/2$ for the heat equation.

This is not an isolated occurrence. General non-negative solutions with compactly supported initial data behave for large times like the ZKB solution if $m > m_c$, and the behaviour of the trajectories is asymptotically similar. Therefore, we know the rate at which particle trajectories go to infinity, which is power-like in time.

The case of exponents $m < m_c$ is discussed in Section 5.5 in connection with escape to infinity in finite time.

Finally, let us point out the surprising fact that formula (AII.6) is exact also for the Barenblatt solutions of the doubly nonlinear equation (11.3).

[1]When $m \geq 1$ a solution has a linear relation x-$\mathbf{v}$ if and only it is a ZKB or its blow-up modifications; for $m_c < m < 1$ we get the solutions described in Exercise 5.2, in particular the singular solution with $C = 0$. For $m = m_c$ we get the non-integrable solution (5.36).

AII.3 Speed distributions

We devote this section to introduce the concept of velocity distribution and derive a curious property of the ZKB solutions.

When we view a solution $u(x,t)$ of the PME/FDE as a mass density, we can think of the mass distribution parametrized by time, which is a family of measures μ_t, $t > 0$, such that for all measurable sets $A \subset \mathbf{R}^n$ we have

$$\mu_t(A) = \int_A u(x,t)\,dx.$$

If the velocity $\mathbf{v}$ is given in terms of x by a bijective C^1 mapping, $x = T_t(v)$ for every $t > 0$, then the change of variables formula implies that for every subset B of velocity values the amount of particles included in that set is

$$\int_{T_t(B)} u(x,t)\,dx = \int_B u(T_t(v),t)|DT_t(v)|\,dv.$$

This means that the density function for the (time family of) velocity distributions is

$$\rho(v,t) = u(T_t v, t)|DT_t(v)|,$$

where the last factor means the absolute value of the determinant of the Jacobian matrix of $T(v,t)$ with respect to $v \in \mathbf{R}^n$.

In the case of the ZKB solutions, we know that $T_t(v) = vt/\beta$, hence $|DT_t(v)| = (t/\beta)^n$ and (putting $\gamma = 1/\beta > 0$) we get

$$\rho(v,t) = B_m(\gamma v t, t)(\gamma t)^n,$$

where B_m is the ZKB solution (2.2). Using the scaling properties of the ZKB solution, this formula can be transformed into

$$\rho(v,t) = (\gamma t)^n t^{-n\beta} F_m(\gamma v t/t^\beta) = (\gamma t^{1-\beta})^n F(v\,\gamma t^{1-\beta}) = B_m(v,t'),$$

where $t' = \beta t^{\beta-1}$. Note that $t' \to 0$ as $t \to \infty$ if $m \geq 1$. It is even true for $\beta < 1$, i.e., for $m > (n-1)/n$. We conclude that

Proposition AII.1 *The distribution density of a ZKB solution in velocity space at time $t > 0$ is again a ZKB solution of the same mass. If $m > (n-1)/n$, it runs in inverse time, $t' = \beta t^{\beta-1}$, so that as $t \to \infty$, this velocity distribution tends to a Dirac delta. On the contrary, for $m_c < m < (n-1)/n$ it runs in positive time towards a distribution with many particles having large speeds. Finally, for the exact exponent $m = (n-1)/n$, the velocity distribution is constant in time.*

The first instance of the result reflects in a quantitative way the fact that most particles have very small velocities as $t \to \infty$ in linear or 'normal' nonlinear diffusion The last part points out the peculiar features of the range of exponents close to m_c, see also [CaVa03].

Comments The Lagrangian formulation of the PME is described in several references, like [GMS] and [SV96]. See the monograph [MPS] for the topic of Eulerian versus Lagrangian systems of coordinates in continuum mechanics.

The special properties of the velocities of ZKB solutions described above seem to be new. I thank A. Bobylev for an interesting discussion of that issue and similar results in kinetic equations.

Appendix III. Some Riemannian geometry

A Riemannian manifold (M, g) is a differentiable (finite-dimensional) manifold M with a Riemannian structure given by a twice-covariant tensor g which is symmetric and positive and allows us to define the length of tangent vectors v as $(g(v, v))^{1/2}$. We will be interested in conformal maps between metrics: we say that g is a metric in the conformal class of g_0 if the matrices representing g and g_0 in some coordinate frame are proportional at all points. The proportionality constant is called the conformal factor.

AIII.1 The Yamabe problem

The Yamabe problem starts with a Riemannian manifold (M, g_0) in space dimension $n \geq 3$ and deals with the question of finding another metric g in the conformal class of g_0 having constant scalar curvature. The formulation of the problem proceeds as follows. We can write the conformal relation as

$$g = u^{4/(n-2)} g_0$$

locally on M for some positive smooth function u. The conformal factor is $u^{4/(n-2)}$. The volume element is

$$|det\, g|^{1/2} dx = u^{2n/(n-2)} dx.$$

The length element is $ds = g^{1/2} dr = u^{2/(n-2)} dr$ where dr is the length element in g_0.

Next, we denote by $R = R_g$ and R_0 the scalar curvatures of the metrics g, g_0, respectively. If we write Δ_0 for the Laplace–Beltrami operator of g_0, we have the formula

$$R = -u^{-N} Lu \quad \text{on } M,$$

with $N = (n + 2)/(n - 2)$ and

$$Lu := \kappa \Delta_0 u - R_0 u, \quad \kappa = \frac{4(n-1)}{n-2}.$$

The Yamabe problem becomes then

$$\Delta_0 u - \left(\frac{n-2}{4(n-1)}\right) R_0 u + \left(\frac{n-2}{4(n-1)}\right) R_g u^{(n+2)/(n-2)} = 0. \qquad \text{(AIII.1)}$$

The equation should determine u (hence, g) when g_0, R_0 and R_g are known. In the standard case we take $M = \mathbf{R}^n$ and g_0 the standard metric, so that Δ_0 is the standard

Laplacian, $R_0 = 0$, we take $R_g = 1$ and then we get the well-known semilinear elliptic equation with critical exponent.

AIII.2 The Yamabe flow and fast diffusion

This is a different but related story. The Yamabe flow is defined as an evolution equation for a family of metrics that is used as a tool to construct metrics of constant scalar curvature within a given conformal class. More precisely, we look for a one-parameter family $g_t(x) = g(x,t)$ of metrics solution of the evolution problem

$$\partial_t g = -R\,g, \quad g(0) = g_0 \quad \text{on } M. \tag{AIII.2}$$

It is easy to show that this is equivalent to the equation

$$\partial_t u^N = Lu, \quad u(0) = 1 \quad \text{on } M,$$

after rescaling the time variable. Let now (M, g_0) be $\mathbf{R}^n$ with the standard flat metric, so that $R_0 = 0$. Put $u^N = v$, $m = 1/N = (n-2)/(n+2) \in (0,1)$. Then

$$\partial_t v = L v^m, \tag{AIII.3}$$

which is a fast diffusion equation with exponent $0 < m < 1$ given by

$$m = \frac{n-2}{n+2}, \qquad 1 - m = \frac{4}{n+2}, \qquad 1 + m = \frac{2n}{n+2}.$$

If we now try separate variables solutions of the form

$$v(x,t) = (T-t)^{\alpha} f(x),$$

then necessarily $\alpha = 1/(1-m) = (n+2)/4$, and $F = f^m$ satisfies the semilinear elliptic equation with critical exponent found before:

$$\Delta F + \frac{n+2}{4} F^{\frac{n+2}{n-2}} = 0. \tag{AIII.4}$$

AIII.3 Two-dimensional Ricci flow. Gauss–Bonnet formula

Now the space dimension is $n = 2$. We write $g = u\,g_0$ where g_0 is the standard metric in $\mathbf{R}^2$ with coordinates (x, y) and u is the conformal factor. Let us put $u = e^{2p}$, $p = p(x,y)$. The length elements in both metrics are related by $ds^2 = e^{2p} dr^2$, where $dr^2 = dx^2 + dy^2$. The Ricci flow has the same formula as for the Yamabe flow:

$$\partial_t g = -R\,g, \quad g(0) = g_0 \quad \text{on } \mathbf{R}^2 \tag{AIII.5}$$

with $R = 2K$. It is shown that the Gaussian curvature K is given by

$$K = -e^{-2p}\,\Delta p = -\frac{\Delta \log u}{2u}.$$

Therefore, the flow is equivalent to $\partial_t(e^{2p}) = 2\Delta p$, hence

$$u_t = \Delta \log u. \tag{AIII.6}$$

We have mentioned in the theory of Chapter 9 the famous Gauss–Bonnet Theorem. This basic result in Riemannian geometry applies to a closed oriented two-dimensional manifold without boundary and says that the following topological restriction holds on the integral of the curvature

$$\int_M K\, dv_g = 2\pi\, \chi(M),$$

where $\chi(M)$ denotes the Euler characteristic of the manifold. In view of our conformal situation we have $dv_g = e^{2p} dx_1 dx_2 = u\, dx$, so that

$$\int_{\mathbf{R}^2} u_t\, dx = 2 \int_{\mathbf{R}^2} \Delta p\, dx = -2 \int_M K\, dv_g.$$

Let us also remark that the total area of the surface at time t is given by

$$A(t) = \int_{\mathbf{R}^2} u(x,t)\, dx_1 dx_2. \tag{AIII.7}$$

A more general Gauss–Bonnet formula deals with an embedded triangle T in any two-dimensional Riemannian manifold M and says that the integral over the whole triangle of the Gaussian curvature with respect to area is given by minus the sum of the jump angles minus the integral of the geodesic curvature over the whole of the boundary of the triangle (with respect to arc length),

$$\int_T K\, dA = 2\pi - \sum_i \alpha_i - \int_{\partial T} \kappa_g\, ds. \tag{AIII.8}$$

cf. [Wol]. This limit as the triangle covers the plane is used in obtaining the formula for the decay rate of maximal solutions of the log-diffusion equation.

AIII.4 Comments and historical notes

The Yamabe problem was posed by H. Yamabe in 1960 [Ya60]. The answer is positive but it took many years to develop. Yamabe attempted to show that any Riemannian metric on a compact manifold can be conformally deformed into one with constant scalar curvature. N. Trudinger discovered an error in Yamabe's proof and was able to solve the problem under some restrictive assumptions. After work by Th. Aubin [Au76b], it was eventually proved by R. Schoen [Sc84] completely. A first survey can be found in [LP87]. See more on the issue in [SY94]. The version for surfaces is the Ricci flow introduced by R. Hamilton in [Ha88], 1988.

Appendix IV. Some extensions and parallel topics

Other equations A very natural extension of the contents of this text is the consideration of the PME/FDE equation with a forcing term, $u_t = \Delta(u^m/m) + f$, where either $f = f(x,t)$ or $f = f(u)$. Many partial results are known, but a systematic study is missing.

We have started the presentation by a more general model than the PME, namely the filtration equation, which includes quite popular applied models like Stefan equation of phase change. There are a number of results on estimates in the vein of what has been explained here.

As an extension of the p-Laplacian equation we could consider the mean curvature evolution equation, where $\Delta_p(u)$ is replaced by $\nabla \cdot ((|1 + |\nabla u|^2)^{-1/2}\nabla u)$, and other models. In particular, our results have corollaries and extensions for such equations that could be interesting to develop.

Convergence to definite profiles in evolution equations as $t \to \infty$ The Solutions of the Cauchy problem for the PME, the FDE, and the PLE not only decay with a precise rate as $t \to \infty$; it can also be proved that generic solutions approach the Barenblatt solution with the same mass as $t \to \infty$. A complete study for the PME is done in [Va03] where further references can be found. For the PLE, cf. [KV88].

Local estimates Many questions remain open. The topic is to be developed separately.

Signed solutions The maximum principle allows us to translate our a priori estimates for non-negative solutions into estimates for solutions of any sign. Note that the PME has to be written then as $u_t = \nabla \cdot (|u|^{m-1}\nabla u)$. Standard theory allows us to construct a semigroup for data in $L^1(\mathbf{R}^n)$ for all $m > 0$, and the data can be extended more or less like in the non-negative case, cf. [V06].

But some questions arise: How sharp are the estimates for particular classes of signed solutions? What is the actual large-time behaviour of such solutions? An example of that behaviour is given in [KV91], where solutions of the PME with data of two signs are shown to become one-signed in finite time (assumptions: u_0 is compactly supported and $\int u_0\,dx \neq 0$; the last condition is essential).

A quite interesting problem is the extension of the theory of non-existence for superfast diffusion to solutions of any sign. Here is a first result obtained by Rodriguez and the author [RV02]: Let $n = 1$ and $m \leq 0$. Then there exists no weak solution of the FDE equation $u_t = (|u|^{m-1}u_x)_x$ having two signs. If we take two-signed data and approximate the problem by regularized solutions, then an initial layer of discontinuity appears. Similar work in several dimensions is needed.

Asymptotic estimates for evolution equations in bounded domains The decay of solutions of parabolic equations in bounded domains with homogeneous Dirichlet data has been studied by different authors using comparison techniques based on symmetrization, like [ALT86, ATL90, Di92, Ta93] among others.

In the case of the PME, the method of symmetrization produces comparison with radial cases, but does not seem to be needed to obtain optimal rates. See more in the survey [Va04] and in [V06].

Fast diffusion equations In work with E. Chasseigne [ChV02], we have studied the fast diffusion equation in the range $m_c < m < 1$, and shown that it admits a well-posed theory of singular solutions with so-called strong singularities. There is even a model singular solution of the form

$$U(x,t) = c_{m,n}\left(t/|x|^2\right)^{\frac{2}{1-m}}.$$

The reader will notice that this solution corresponds to the explicit extinction model of the range $m < m_c$ used in Chapter 5. We wonder how to apply the worst case strategy in that framework, though some work in that direction is done in our mentioned paper.

Bibliography

[AP04] B. ABDELLAOUI, I. PERAL. Hölder regularity and Harnack inequality for degenerate parabolic equations related to Caffarelli–Kohn–Nirenberg inequalities, *Nonlinear Anal.* **57**, no. 7–8 (2004), 971–1003.

[Ad75] R. ADAMS. *Sobolev Spaces*, Pure and Applied Mathematics, Vol. 65. Academic Press, New York, 1975.

[AR81] N. D. ALIKAKOS, R. ROSTAMIAN. Large time behavior of solutions of Neumann boundary value problem for the porous medium equation, *Indiana Univ. Math. J.* **30** (1981), no. 5, 749–785.

[ALT86] A. ALVINO, P. L. LIONS, G. TROMBETTI. Comparaison des solutions d'équations paraboliques et elliptiques par symétrisation. Le cas de la symétrisation de Schwarz, *Comptes Rendus Acad. Sci. Paris, Série I* **303** (1986), 947–950 (in French).

[ATL90] A. ALVINO, G. TROMBETTI, P. L. LIONS. Comparison results for elliptic and parabolic equations via Schwarz symmetrization, *Ann. Inst. H. Poincaré, Anal. Non Linéaire* **7** (1990), no. 2, 37–65.

[An04] M. T. ANDERSON. Geometrization of 3-manifolds via the Ricci flow, *Notices Amer. Math. Soc.* **51** (2004), no. 2, 184–193.

[ACM04] A. ANDREU-VAILLO, V. CASELLES, J. M. MAZÓN. *Parabolic Quasilinear Equations Minimizing Linear Growth Functionals*, Birkhäuser, Basel, 2004.

[AA95] S. B. ANGENENT, D. G. ARONSON. The focusing problem for the radially symmetric porous medium equation, *Comm. Partial Diff. Eqns.* **20** (1995), 1217–1240.

[AA01] S. B. ANGENENT, D. G. ARONSON. Non-axial self-similar hole filling for the porous medium equation, *J. Amer. Math. Soc.* **14** (2001), no. 4, 737–782.

[Ar67] D. G. ARONSON. Bounds on the fundamental solution of a parabolic equation, *Bull. Amer. Math. Soc.* **73** (1967), 890–896.

[Ar86] D. G. ARONSON. The porous medium equation. In *Nonlinear Diffusion Problems*, Lecture Notes in Math. **1224**, A. Fasano and M. Primicerio eds., Springer-Verlag New York, 1986, pp. 12–46.

[AB79] D. G. ARONSON, PH. BÉNILAN. Régularité des solutions de l'équation des milieux poreux dans R^n, *C. R. Acad. Sci. Paris Ser. A-B* **288** (1979), 103–105 (in French).

[ABH03] D. G. ARONSON, J. B. VAN DEN BERG, J. HULSHOF. Parametric dependence of exponents and eigenvalues in focussing porous media flows, *European J. Appl. Math.* **14** (2003), no. 4, 485–512.

[AC83] D. G. Aronson, L. A. Caffarelli. The initial trace of a solution of the porous medium equation, *Trans. Amer. Math. Soc.* **280** (1983), 351–366.

[AGV98] D. G. Aronson, O. Gil, J. L. Vázquez. Limit behaviour of focusing solutions to nonlinear diffusions, *Comm. Partial Diff. Eqns* **23** (1998), no. 1–2, 307–332.

[AG93] D. G. Aronson, J. A. Graveleau. Self-similar solution to the focusing problem for the porous medium equation, *European J. Appl. Math.* **4** (1993), no. 1, 65–81.

[AV87] D. G. Aronson, J. L. Vázquez. The porous medium equation as a finite-speed approximation to a Hamilton–Jacobi equation, *Ann. Inst. H. Poincaré Anal. Non Linéaire* **4** (1987), no. 3, 203–230.

[AV95] D. G. Aronson, J. L. Vázquez. Anomalous exponents in nonlinear diffusion, *J. Nonlinear Science* **5**, 1 (1995), 29–56.

[Au76b] T. Aubin. The scalar curvature. In *Differential Geometry and Relativity*, edited by Cahen and Flato, Reidel, Dordrecht, 1976, pp. 5–18.

[Au76] T. Aubin. Problèmes isopérimétriques et espaces de Sobolev, *J. Differential Geometry* **11**, no. 4 (1976), 573–598.

[Bk94] D. Bakry. L'hypercontractivité et son utilisation en théorie des semigroupes. Ecole d'été de Probabilités de St-Flour, *Lecture Notes in Math.* **1581**, pp. 1–114, (1994). Springer (in French).

[BMP02] A. Baldassarri, U. Marini B. M., A. Puglisi. Cooling of a lattice granular fluid as an ordering process, *Phys. Rev. E* **65**, (2002) 051301.

[B80] C. Bandle. *Isoperimetric Inequalities and Applications*, Pitman, Boston, 1980.

[Ba52] G. I. Barenblatt. On some unsteady motions of a liquid or a gas in a porous medium, *Prikl. Mat. Mekh.* **16** (1952), 67–78 (in Russian).

[Ba52b] G. I. Barenblatt. On self-similar motions of compressible fluids in porous media, *Prikl. Mt. Makh.* **16**, 6 (1952), 679–698 (in Russian).

[Ba79] G. I. Barenblatt. *Scaling, Self-Similarity, and Intermediate Asymptotics*, Cambridge University Press, Cambridge, 1996. Updated version of *Similarity, Self-Similarity, and Intermediate Asymptotics*, Consultants Bureau, New York, 1979.

[Ba87] G. I. Barenblatt. *Dimensional Analysis*, Gordon and Breach, New York, 1987.

[BER90] G. I. Barenblatt, V. M. Entov, V. M. Ryzhik. *Flow of Fluids Through Natural Rocks*, Kluwer Academic, 1990.

[BS69] G. I. Barenblat, G. I. Sivashinsky. Self-similar solutions of the second type in nonlinear filtration, *Appl. Math. Mech.* **33**, no. 5 (1969), 861–870 (Russian); translated as J. Appl. Math. Mech. 33 1969, 836–845 (1970).

[BV04] G. I. Barenblatt, J. L. Vázquez. Nonlinear diffusion and image contour enhancement, *Interfaces Free Boundaries* **6** (2004), 31–54.

[Bea72] J. Bear. *Dynamics of Fluids in Porous Media*, Dover, New York, 1972.

[BE89] J. Bebernes, D. Eberly. *Mathematical Problems from Combustion Theory*, Appl. Math. Sci. **83**, Springer-Verlag, New York, 1989.

[Be72] P. BÉNILAN. *Equations d'évolution dans un espace de Banach quelconque et applications*, Thèse, Orsay, 1972.

[Be76] PH. BÉNILAN. Opérateurs accrétifs et semi-groupes dans les espaces L^p $(1 \leq p \leq \infty)$, *France-Japan Seminar,* Tokyo, 1976.

[Be83] P. BÉNILAN. A strong regularity L^p for solutions of the porous media equation, *Research Notes in Math.* **89**, 39–58, Pitman, London, 1983.

[BB85] P. BÉNILAN, J. BERGER. Estimation uniforme de la solution de $u_t = \Delta\phi(u)$ et caractérisation de l'effet régularisant, *Comptes Rendus Acad. Sci. Paris* **300**, Série I (1985), 573–576.

[B6-95] P. BÉNILAN, L. BOCCARDO, T. GALLOUËT, R. GARIEPY, M. PIERRE, J. L. VÁZQUEZ. An L^1 theory of existence and uniqueness of solutions of nonlinear elliptic equations, *Annali Scuola Normale Sup. Pisa*, Serie IV, **12**, 2 (1995), 241–273.

[BBH] P. BÉNILAN, L. BOCCARDO, M. A. HERRERO. On the limit of solutions of $u_t = \Delta u^m$ as $m \to \infty$, *Rend. Sem. Mat. Univ. Politec. Torino* (1989), Fascicolo Speciale, 1–13.

[BBC75] P. BÉNILAN, H. BREZIS, M. G. CRANDALL. A Semilinear Equation in $L^1(\mathbf{R}^n)$, *Ann. Scuola Norm. Sup. Pisa* **2** (1975), 523–555.

[BC81] P. BÉNILAN, M. G. CRANDALL. The continuous dependence on φ of solutions of $u_t - \Delta\varphi(u) = 0$. *Indiana Univ. Math. J.* **30** (1981), 161–177

[BC81b] P. BÉNILAN, M. G. CRANDALL. Regularizing effects of homogeneous evolution equations. In *Contributions to Analysis and Geometry*, Suppl. to Amer. Jour. Math., Baltimore (1981), 23–39.

[BCP] P. BÉNILAN, M. G. CRANDALL, A. PAZY. *Evolution Equations Governed by Accretive Operators*, unpublished book draft.

[BEG] P. BÉNILAN, L. C. EVANS, R. F. GARIEPY. On some singular limits of homogeneous semigroups, *J. Evol. Eqns.* **3** (2003), no. 2, 203–214.

[BS88] C. BENNETT, R. SHARPLEY. *Interpolation of Operators*, Pure and Applied Mathematics **129**, Academic Press, Boston, MA, 1988.

[BLP81] H. BERESTYCKI, P. L. LIONS, L. A. PELETIER. An ODE approach to the existence of positive solutions for semilinear problems in R^N, *Indiana Univ. Math. J.* **30** (1981), no. 1, 141–157.

[Bd01] G. BERNARD. Existence theorems for fast diffusion equations, *Nonlinear Anal.* **43** (2001), no. 5, Ser. A: Theory Methods, 575–590.

[BH80] J. B. BERRYMAN, CH. J. HOLLAND. Stability of the separable solution for fast diffusion, *Arch. Rat. Mech. Anal.* **74** (1980), 379–388.

[BHi91] M. BERTSCH, D. HILHORST. The interface between regions where $u < 0$ and $u > 0$ in the porous medium equation, *Appl. Anal.* **41** (1991), no. 1–4, 111–130.

[BDU90] M. BERTSCH, R. DAL PASSO, M. UGHI. Discontinuous "viscosity" solutions of a degenerate parabolic equation, *Trans. Amer. Math. Soc.* **320** (1990), 779–798.

[BDU92] M. BERTSCH, R. DAL PASSO, M. UGHI. Nonuniqueness of solutions of a degenerate parabolic equation, *Annali Mat. Pura Appl.* **161** (1992), 57–81.

[BU90] M. BERTSCH, M. UGHI. Positivity properties of viscosity solutions of a degenerate parabolic equation. *Nonlinear Anal. TMA* **14** (1990), no. 7, 571–592.

[BiV05] M.F. BIDAUT-VERON. Self-similar solutions of the p-Laplace heat equation: the fast diffusion case, *to appear.*

[BK80] G. BLUMAN, S. KUMEI. On the remarkable nonlinear diffusion equation $\partial/\partial x[a(u+b)^{-2}(\partial u/\partial x)] - (\partial u/\partial t) = 0$, *J. Math. Phys.* **21** (1980), 1019–1023.

[BG05] M. BONFORTE, G. GRILLO. Asymptotics of the porous media equation via Sobolev inequalities, *J Functional Anal.* **225**, Issue 1 (2005), 33–62.

[BV05] M. BONFORTE, J. L. VÁZQUEZ. Global positivity estimates and Harnack inequalities for the fast diffusion equation, *preprint, 2005.*

[BV06] M. BONFORTE, J. L. VÁZQUEZ. Fine asymptotics near extinction and elliptic Harnack inequalities for the fast diffusion equation, *preprint, 2006.*

[Bo03] J. BOUSSINESQ. Recherches théoriques sur l'écoulement des nappes d'eau infiltrés dans le sol et sur le débit de sources, *Comptes Rendus Acad. Sci./J. Math. Pures Appl.* **10** (1903/04), pp. 5–78.

[Br71] H. BREZIS. Propriétés régularisantes de certains semi-groupes non linéaires, *Israel J. Math.* **9** (1971), 513–534.

[Br71b] H. BREZIS. Monotonicity methods in Hilbert spaces and some applications to nonlinear partial differential equations, in *Proc. Symp. Nonlinear Funct. Anal.*, Madison (1971), Contributions to Nonlinear Funct. Analysis. Academic Press, pp. 101–156.

[Br73] H. BREZIS. *Opérateurs maximaux monotones et semigroupes de contractions dans les espaces de Banach*, North Holland, Amsterdan 1973.

[BrC96] H. BREZIS, T. CAZENAVE. A nonlinear heat equation with singular initial data, *J. Analyse Math.* **68** (1996), 277–304.

[BrC79] H. BREZIS, M. CRANDALL. Uniqueness of solutions of the initial value problem for $u_t - \Delta\varphi(u) = 0$, *J. Math. Pures Appl.* **58** (1979), 153–163.

[BF83] H. BREZIS, A. FRIEDMAN. Nonlinear parabolic equations involving measures as initial conditions, *J. Math. Pures Appl.* **62** (1983), 73–97.

[BN83] H. BREZIS, L. NIRENBERG. Positive solutions of nonlinear elliptic equations involving critical exponents, *Comm. Pure Appl. Math* **36** (1983), 437–477.

[CF79] L.A. CAFFARELLI, A. FRIEDMAN. Continuity of the density of a gas flow in a porous medium, *Trans. Amer. Math. Soc.* 252 (1979), 99–113.

[CF87] L.A. CAFFARELLI, A. FRIEDMAN. Asymptotic behaviour of solutions of $u_t = \Delta u^m$ as $m \to \infty$, *Indiana Univ. Math. Jour.* **36** (1987), no. 4, 711–728.

[CVW87] L.A. CAFFARELLI, J. L. VÁZQUEZ, N. I. WOLANSKI. Lipschitz continuity of solutions and interfaces of the N-dimensional porous medium equation, *Indiana Univ. Math. J.* **36** (1987), 373–401.

[C5-02] N. Calvo, J.I. Díaz, J. Durany, E. Schiavi, C. Vázquez. On a doubly nonlinear parabolic obstacle problem modelling ice sheet dynamics, *SIAM J. Appl. Math.* **63** (2002), 683–707.

[Cl57] T. Carleman. Problèmes mathématiques dans la thèorie cinétique des gaz, Almqvist-Wiksells, Uppsala, 1957.

[CT00] J. A. Carrillo, G. Toscani. Asymptotic L^1-decay of solutions of the porous medium equation to self-similarity. *Indiana Univ. Math. J.* **49** (2000), 113–141.

[CJMU] J. A. Carrillo, A. Jüngel, P. A. Markowich, G. Toscani, A. Unterreiter. Entropy dissipation methods for degenerate parabolic systems and generalized Sobolev inequalities. *Monatsh. Math.* **133** (2001), 1–82.

[CaVa03] J. A. Carrillo, J. L. Vázquez. Fine asymptotics for fast diffusion equations, *Comm. Partial Differential Eqns* **28** (2003), no. 5–6, 1023–1056.

[ChV02] E. Chasseigne, J. L. Vázquez. Theory of extended solutions for fast diffusion equations in optimal classes of data. Radiation from singularities, *Archive Rat. Mech. Anal.* **164** (2002), 133–187.

[ChV02b] E. Chasseigne, J. L. Vázquez. Sets of admissible initial data for porous medium equations with absorption. *Electron. J. Diff. Eqns., Conf.* **08** (2002), 53–83 (*2001-Luminy Conference on Quasilinear Elliptic and Parabolic Equations and Systems*).

[ChOR93] J. T. Chayes, S. J. Osher, J. V. Ralston. On singular diffusion equations with applications to self-organized criticality, *Comm. Pure Appl. Math.* **46** (1993), no. 10, 1363–1377.

[ChS84] F. M. Chiarenza, R. P. Serapioni. A Harnack inequality for degenerate parabolic equations, *Commun. Partial Diff. Equns.* **9** (8) (1984) 719–749.

[Ch91] B. Chow. The Ricci flow on the 2-sphere, *J. Differential Geom.* **33** (1991), no. 2, 325–334.

[CL71] M. G. Crandall, T. M. Liggett. Generation of semi-groups of nonlinear transformations on general Banach spaces, *Amer. J. Math.* **93** (1971) 265–298.

[DK88] B. E. J. Dahlberg, C. Kenig. Nonnegative solutions to fast diffusion equations, *Rev. Mat. Iberoamericana* **4** (1988), 11–29.

[D4-97] G. Dal Masso, F. Murat, L. Orsina, A. Prignet. Definition and existence of renormalized solution of elliptic equations with general measure data. *C. R. Acad. Sci. Paris Série I Math.* **325**, 5 (1997), 481–486.

[DL87] R. Dal Passo, S. Luckhaus. A degenerate diffusion problem not in divergence form, *J. Differential Eqns* **69** (1987), no. 1, 1–14.

[DH04] P. Daskalopoulos, R. Hamilton. Geometric estimates for the logarithmic fast diffusion equation. *Comm. Anal. Geom.* **12** (2004), no. 1–2, 143–164.

[DP95] P. Daskalopoulos, M. del Pino. On a singular diffusion equation, *Comm. Analysis and Geometry* **3**, 1995, pp 523–542.

[DP97] P. DASKALOPOULOS, M. DEL PINO. On nonlinear parabolic equations of very fast diffusion, *Arch. Rational Mech. Anal.* **137** (1997), no. 4, 363–380.

[DP99] P. DASKALOPOULOS, M. DEL PINO. On the Cauchy problem for $u_t = \Delta \log u$ in higher dimensions, *Math. Ann.* **313** (1999), no. 2, 189–206.

[DP04] P. DASKALOPOULOS, M. DEL PINO. Geometric estimates for the logarithmic fast diffusion equation, *Comm. Anal. Geom.* **12**, Number 1 (2004), 143–164.

[Dv89] E. B. DAVIES. *Heat Kernels and Spectral Theory*. Cambridge University Press, 1989.

[DG57] E. DE GIORGI. Sulla differenziabilitá e analiticitá delle estremali degli integrali multipli regolari, *Mem. Accad. Sci. Torino Cl. Sci.* **3** (1957) 25–43 (in Italian).

[DPD02] M. DEL PINO, J. DOLBEAULT. Best constants for Gagliardo–Nirenberg inequalities and applications to nonlinear diffusions, *J. Math. Pures Appl.* **9**, 81 (2002), pp. 847–875.

[DPD02b] M. DEL PINO, J. DOLBEAULT. Nonlinear diffusions and optimal constants in Sobolev type inequalities: asymptotic behaviour of equations involving the p-Laplacian, *C. R. Math. Acad. Sci. Paris* **334** (2002), 365–370.

[DDG04] M. DEL PINO, J. DOLBEAULT, I. GENTIL. Nonlinear diffusions, hypercontractivity and the optimal L^p-Euclidean logarithmic Sobolev inequality, *J. Math. Anal. Appl.* **293** (2004), 375–388.

[DPS01] M. DEL PINO, M. SAEZ. On the extinction profile for solutions of $u_t = \Delta u^{(N-2)/(N+2)}$, *Indiana Univ. Math. Journal* **50**, 2 (2001), 612–628.

[De05] J. DEMANGE. *Des équations à diffusion rapide aux inégalités de Sobolev sur les modèles de la géométrie,* Ph. D. Thesis, 2005, Univ. Toulouse (in French).

[DD79] G. DIAZ, J. I. DIAZ. Finite extinction time for a class of non-linear parabolic equations, *Comm. Partial Diff. Eqns.* **4** (1979), 1213–1231.

[Di92] J. I. DIAZ. Symmetrization of nonlinear elliptic and parabolic problems and applications: a particular overview, *Proc. European Conf. on Elliptic and Parabolic Problems*, Progress in partial differential equations: elliptic and parabolic problems (Pont-à-Mousson, 1991), 1–16, Pitman Res. Notes Math. Ser., 266, Longman Harlow, 1992.

[DiB88] E. DIBENEDETTO. Intrinsic Harnack type inequalities for solutions of certain degenerate parabolic equations, *Arch. Rational Mech. Anal.* **100** (1988), no. 2, 129–147.

[DiB93] E. DIBENEDETTO. *Degenerate Parabolic Equations,* Series Universitext, Springer-Verlag, New York, 1993.

[DBD96] E. DIBENEDETTO, D. J. DILLER. About a singular parabolic equation arising in thin film dynamics and in the Ricci flow for complete R^2, *Partial Differential Equations and Applications*, 103–119, Lecture Notes in Pure and Appl. Math., **177**, Dekker, New York, 1996.

[DF85] E. DIBENEDETTO, A. FRIEDMAN. Hölder estimates for nonlinear degenerate parabolic systems, *J. Reine Angew. Math.* **357** (1985), 1–22. Addendum to: "Hölder estimates for nonlinear degenerate parabolic systems", *J. Reine Angew. Math.* **363** (1985), 217–220.

[DH89] E. DIBENEDETTO, M. A. HERRERO. On the Cauchy problem and initial trace for a degenerate parabolic equation, *Trans. Am. Math. Soc.* **314** (1989), 187–224.

[DH90] E. DIBENEDETTO, M. A. HERRERO. Nonnegative solutions of the evolution p-Laplacian equation, initial traces and Cauchy problem when $1 < p < 2$, *Arc. Rat. Mech. Anal* **11** (1990), 225–290.

[DK92] E. DIBENEDETTO, Y. C. KWONG. Harnack estimates and extinction profile for weak solutions of certain parabolic equations, *Trans. Am. Math. Soc.* **330** (1992) 783–811.

[DKV91] E. DIBENEDETTO, Y. C. KWONG, V. VESPRI. Local space analyticity and asymptotic behavior of solutions of certain singular parabolic equations, *Indiana Univ. Math. J.* **40** (1991), 741–765.

[E4-86] C. M. ELLIOTT, M. A. HERRERO, J. R. KING, J. R. OCKENDON. The mesa problem: diffusion patterns for $u_t = \nabla \cdot (u^{m-1}\nabla u)$ as $m \to \infty$, *IMA J. Appl. Math.* **37** (1986), 147–154.

[ERV88] J. R. ESTEBAN, A. RODRÍGUEZ, J. L. VÁZQUEZ. A nonlinear heat equation with singular diffusivity, *Comm. Partial Diff. Eqs.* **13** (1988), 985–1039.

[EV86] J. R. ESTEBAN, J. L. VÁZQUEZ. On the equation of turbulent filtration in one-dimensional porous media, *Nonlinear Anal. TMA* **10** (1986), 1303–1325.

[EV88] J. R. ESTEBAN, J. L. VÁZQUEZ. Homogeneous diffusion in **R** with power-like nonlinear diffusivity, *Archive Rat. Mech. Anal.* **103** (1988), 39–80.

[Ev78] L. C. EVANS. Applications of nonlinear semigroup theory to certain partial differential equations, in *Nonlinear Evolution Equations*, M. G. Crandall ed., Academic Press, New York 1978, pp. 163–188.

[EFG97] L. C. EVANS, M. FELDMAN, R. F. GARIEPY. Fast/slow diffusion and collapsing sandpiles, *J. Differential Eqns* **137** (1997), no. 1, 166–209.

[FV03] R. FERREIRA, J. L. VÁZQUEZ. Study of self-similarity for the fast diffusion equation, *Advances Diff. Eqns.* **8**, no 9 (2003), 1125–1152.

[Fk00] L. E. FRAENKEL. *An Introduction to Maximum Principles and Symmetry in Elliptic Problems.* Cambridge Tracts in Mathematics, **128.** Cambridge University Press, Cambridge, 2000.

[F64] A. FRIEDMAN. *Partial Differential Equations of Parabolic Type*, Prentice-Hall, Englewood Cliffs, NJ, 1964.

[FH] A. FRIEDMAN, S. HUANG. Asymptotic behaviour of solutions of $u_t = \Delta\varphi_m(u)$ as $m \to \infty$ with inconsistent initial values, *Analyse mathématique et applications*, 165–180, Gauthier-Villars, Montrouge, 1988.

[FK80] A. FRIEDMAN, S. KAMIN. The asymptotic behaviour of gas in an n-dimensional porous medium, *Trans. Amer. Math. Soc.* **262** (1980), 551–563.

[Ga04] V. A. GALAKTIONOV. *Geometric Sturmian Theory of Nonlinear Parabolic Equations and Applications*, Chapman & Hall/CRC Applied Mathematics and Nonlinear Science Series, **3**. Chapman & Hall/CRC, Boca Raton, FL, 2004.

[GK02] V. A. GALAKTIONOV, J.R. KING. Fast diffusion equation with critical Sobolev exponent in a ball, *Nonlinearity* **15** (2002), no. 1, 173–188.

[GP97] V. A. GALAKTIONOV, L.A. PELETIER. Asymptotic behaviour near finite time extinction for the fast diffusion equation, *Arch. Rational Mech. Anal.* **139** (1997), no. 1, 83–98.

[GPV00] V. A. GALAKTIONOV, L. A. PELETIER, J. L. VÁZQUEZ. Asymptotics of the fast-diffusion equation with critical exponent, *SIAM J. Math. Anal.* **31** 5 (2000), 1157–1174.

[GV91] V. A. GALAKTIONOV, J. L. VÁZQUEZ. Asymptotic behaviour of nonlinear parabolic equations with critical exponents. A dynamical systems approach, *J. Funct. Anal.* **100** (1991), no. 2, 435–462.

[GV97] V. A. GALAKTIONOV, J. L. VÁZQUEZ. Continuation of blowup solutions of nonlinear heat equations in several space dimensions, *Comm. Pure Applied Math.* **1**, 1 (1997), 1–67.

[GV02] V. A. GALAKTIONOV, J. L. VÁZQUEZ. The problem of blow-up in nonlinear parabolic equations, *Discrete Contin. Dynam. Systems A* **8** (2002), 399–433.

[GV03] V.A. GALAKTIONOV, J. L. VÁZQUEZ. *A Stability Technique for Evolution Partial Differential Equations. A Dynamical Systems Approach*. PNLDE 56 (Progress in Non-Linear Differential Equations and Their Applications), Birkhäuser Verlag, 2003.

[Ge84] P. G. DE GENNES. Spreading laws for microscopic droplets, *C.R. Acad. Sci., Paris II,* **298** (1984), 475–478.

[Ge85] P. G. DE GENNES. Wetting: statics and dynamics, *Rev. Mod. Phys.* **57** No 3, (1985), 827–863.

[GQ03] O. GIL, F. QUIROS. Boundary layer formation in the transition from the porous media equation to a Hele–Shaw flow, *Annales de l'Institut Henri Poincare (C) Nonnear Analysis* **20** (2003), no. 1, 13–36.

[GH83] J. GUCKENHEIMER, PH. HOLMES. *Nonlinear Oscillations, Dynamical Systems, and Bifurcations of Vector Fields*, Applied Mathematical Sciences, **42**. Springer-Verlag, New York, 1983, revised edition 1990.

[GMS] M.E. GURTIN, R.C. MCCAMY, E. SOCOLOVSKI. A coordinate transformation for the porous media equation that renders the free boundary stationary, *Quart. Appl. Math.* **42** (1984), no. 3, 345–357.

[Ha88] R. S. HAMILTON. The Ricci flow on surfaces, *Contemporary Math.* **71** (1988), 237–262.

[HLP64] G. M. HARDY, J. E. LITTLEWOOD, G. PÓLYA. *Inequalities*, Cambridge University Press, 1964.

[Hb99] E. HEBEY, *Nonlinear Analysis on Manifolds: Sobolev Spaces and Inequalities*, Courant Lecture Notes in Mathematics, vol. 5, New York University, Courant Institute of Mathematical Sciences, New York, 1999.

[HP58] F. HELFFERICH, M. S. PLESSET. Ion exchange kinetics: A nonlinear diffusion problem, *J. Chem. Phys.* **2**, 3 (1958), 418–424.

[H89] M. A. HERRERO. A limit case in nonlinear diffusion, *Nonlinear Anal.* **13** (1989), 611–628.

[HP85] M. A. HERRERO, M. PIERRE. The Cauchy problem for $u_t = \Delta u^m$ when $0 < m < 1$, *Trans. AMS* **291** (1985), 145–158.

[HV81] M. A. HERRERO, J. L. VÁZQUEZ, Asymptotic behaviour of the solutions of a strongly nonlinear parabolic problem, *Ann. Sci. Toulouse* **3** (1981), 113–127.

[Hs02] S.-Y. HSU. Dynamics near extinction time of a singular diffusion equation, *Math. Ann.* **323** (2002), no. 2, 281–318.

[Hs02b] S.-Y. HSU. Asymptotic profile of solutions of a singular diffusion equation as $t \to \infty$, *Nonlinear Anal. Ser. A: Theory Methods* **48** (2002), no. 6, 781–790.

[Hs03] S.-Y. HSU. Asymptotic behavior of solutions of the equation $u_t = \Delta \log u$ near the extinction time, *Adv. Differential Eqns*, **8**, No 2, (2003), pp 161–187.

[Hs04] S.-Y. HSU. Behaviour of solutions of a singular diffusion equation near the extinction time, *Nonlinear Anal.* **56** (2004), no. 1, 63–104.

[Hs04] S.-Y. HSU. Global existence and uniqueness of solutions of the Ricci flow equation, *Differential Integral Eqns* **14** (2001), no. 3, 305–320.

[Hu94] K. M. HUI. Nonexistence of fundamental solutions of the equation $u_t = \Delta logu$, *J. Math. Anal. Appl.* **182** (1994), no. 3, 800–809.

[Hu99] K. M. HUI. Existence of solutions of the equation $u_t = \Delta \log u$, *Nonlinear Anal. TMA* **37** (1999), 875–914.

[Hu99] K. M. HUI. Singular limit of solutions of the equation $u_t = \Delta(u^m/m)$ as $m \to 0$, *Pacific Jour. Math.* **187**, No. 2 (1999), 297–316.

[HuV91] J. HULSHOF, J. L. VÁZQUEZ. The dipole solution for the porous medium equation in several space dimensions, *Ann. Scuola Norm. Sup. Pisa Cl. Sci.* (4) **20**, no. 2 (1993), 193–217.

[Ht97] K. H. HUTTER. Mathematical foundation of ice sheet and ice shelf dynamics: A physicist's view, *Free Boundary Problems: Theory and Applications de I* (Athanasopoulos, G. Makrakis, Jose-Francisco Rodrigues eds.) (Crete, 1997), 192–203, Chapman & Hall/CRC Res. Notes Math., 409, Boca Raton, FL, 1999.

[JL73] D. D. JOSEPH, T. S. LUNDGREN. Quasilinear Dirichlet problems driven by positive sources, *Arch. Ration. Mech. Anal.* **49** (1973), 241–269.

[Ka87 A. S. KALASHNIKOV. Some problems of the qualitative theory of nonlinear degenerate second-order parabolic equations, *Russian Math. Surveys* **42** (1987), 169–222.

[Km73] S. KAMIN. The asymptotic behaviour of the solution of the filtration equation, *Israel J. Math.* **14** (1973), 76–87.

[KD01] S. KAMIN, L. DASCAL. Long time slow expansion of hot bubbles in gases, *Optimal Control and Partial Differential Equations*, J. L. Menaldi et al. (eds). IOS Press, 2001.

[KV88] S. KAMIN, J. L. VÁZQUEZ. Fundamental solutions and asymptotic behaviour for the p-Laplacian equation, *Revista Mat. Iberoamericana* **4** (1988), 339–354.

[KV91] S. KAMIN, J. L. VÁZQUEZ. Asymptotic behaviour of solutions of the porous medium equation with changing sign, *SIAM J. Math. Anal.* **22** (1991), 34–45.

[Kw95] B. KAWOHL. *Rearrangements and Convexity of Level Sets in PDE*, Lecture Notes in Mathematics, vol. **1150**, Springer Verlag, Berlin, 1895.

[Ki88] J. R. KING. Extremely high concentration dopant diffusion in silicon, *IMA J. Appl. Math.* **40** (1988), 163–181.

[Ki93] J. R. KING. Self-similar behaviour for the equation of fast nonlinear diffusion, *Phil. Trans. Roy. Soc. London A* **343** (1993), 337–375.

[KN02] H. KUHN, S. NASAR. *The Essential John Nash*, Princeton University Press, 2002.

[Ku73] T. G. KURTZ. Convergence of sequences of semigroups of nonlinear operators with an application to gas kinetics, *Trans. Amer. Math. Soc.* **186** (1973), 259–272.

[Kw98] Y. C. KWONG. Asymptotic behaviour of a plasma type equation with finite extinction, *Arch. Ration. Mech. Anal.* **104** (1988), no. 3, 277–294.

[LOT82] A. A. LACEY, J. R. OCKENDON, A. B. TAYLER. "Waiting-time" solutions of a nonlinear diffusion equation, *SIAM J. Appl. Math.* **42** (1982), 1252–1264.

[La69] O. A. LADYZHENSKAYA. *The Mathematical Theory of Viscous Incompressible Flow*, Gordon and Breach, 1969.

[LSU] O. A. LADYZHENSKAYA, V. A. SOLONNIKOV, N. N. URALTSEVA. *Linear and Quasilinear Equations of Parabolic Type*, Amer. Math. Soc., Providence, RI, 1968.

[LP87] J. M. LEE, T. PARKER. The Yamabe problem, *Bull Am. Math. Soc.* **17**, 1 (1987), 37–91.

[Le29] L. S. LEIBENZON. The motion of a gas in a porous medium, *Complete Works*, vol **2**, Acad. Sciences URSS, Moscow, 1953 (in Russian). First published in *Neftanoe i slantsevoe khozyastvo,* **10**, 1929, and *Neftanoe khozyastvo,* **8–9**, 1930 (in Russian).

[Le45] L. S. LEIBENZON. General problem of the movement of a compressible fluid in a porous medium, *Izv. Akad. Nauk SSSR*, Geography and Geophysics **9** (1945), 7–10 (in Russian).

[Lb83] E. LIEB. Sharp constants in the Hardy–Littlewood–Sobolev inequality, *Ann. of Math.* **118** (1983), 345–374.

[LSV87] P. L. LIONS, P. E. SOUGANIDIS, J. L. VÁZQUEZ. The relation between the porous medium and the eikonal equations in several space dimensions, *Rev. Mat. Iberoamericana* **3** (1987), no. 3–4, 275–310.

[LiTo97] P. L. LIONS, G. TOSCANI. Diffusive limits for finite velocities Boltzmann kinetic models, *Rev. Mat. Iberoamericana*, **13** (1997), 473–513.

[LN74] C. LOEWNER, L. NIRENBERG. Partial differential equations invariant under conformal or projective transformations, *Contributions to Analysis*, Academic Press, New York, 1974, pp. 245–272.

[LH76] K. E. LONNGREN, A. HIROSE. Expansion of an electron cloud, *Physics Lett. A*, **59** (4) (1976), 285–286.

[LMR76] J. LÓPEZ, C. A. MILLER, E. RUCKENSTEIN. Spreading kinetics of liquid drops on solids, *J. Coll. Int. Sci.* **56** (1976), 460–461.

[MV94] J. J. MANFREDI, V. VESPRI. Large time behavior of solutions to a class of doubly nonlinear parabolic equations, *Electron. J. Differential Eqns*, 1994, No. 02, approx. 17 pp. (electronic only).

[Ma82] H. MATANO. Nonincrease of the lap number for a one-dimensional semilinear parabolic equation, *J. Fac. Sci. Univ. Tokyo, Sec. IA* **29** (1982), 401–441.

[McK75] H. P. MCKEAN. The central limit theorem for Carleman's equation, *Israel J. Math.* **21**, no. 1 (1975), 54–92.

[Me89] B. MEERSON. On the dynamics of strong temperature disturbances in the upper atmosphere of the Earth, *Phys. Fluids A* **1** (1989), 887–891.

[MSS] B. MEERSON, P.V. SASOROV, K. SEKIMOTO. Logarithmically slow expansion of hot bubbles in gases, *Phys. Rev. E* **61** (2000), 1403–1406.

[MPS] A. M. MEIRMANOV, V. V. PUKHNACHOV, S. I. SHMAREV. *Evolution Equations and Lagrangian Coordinates*, de Gruyter Expositions in Mathematics, 24. Walter de Gruyter, Berlin, 1997.

[Mo61] J. MOSER. On Harnack's theorem for elliptic differential equations, *Comm. Pure and Appl. Math.* **14** (1961) 577–591.

[Mo64] J. MOSER. A Harnack inequality for parabolic differential equations, *Comm. Pure and Appl. Math.* **17** (1964), 101–134.

[Mu37] M. MUSKAT. *The Flow of Homogeneous Fluids through Porous Media*, Mc Graw-Hill, New York, 1937.

[Na58] J. NASH. Continuity of solutions of parabolic and elliptic equations, *Amer. J. Math.* **80** (1958), 931–954.

[Na58] W.-M. NI, P. SACKS. Singular behavior in nonlinear parabolic equations, *Trans. Amer. Math. Soc.* **287** (1985), 657–671.

[OD73] H. OKUDA, J. M. DAWSON. Theory and nunmerical simulation on plasma diffusion across a magnetic field, *Phys. Fluids* **16** (1973), 408–426.

[OKC58] O. A. OLEINIK, A. S. KALASHNIKOV, Y.-I. CHZOU. The Cauchy problem and boundary problems for equations of the type of unsteady filtration, *Izv. Akad. Nauk SSR Ser. Math.* **22** (1958), 667–704.

[Ot01 F. OTTO. The geometry of dissipative evolution equations: the porous medium equation, *Comm. Partial Differential Eqns* **26** (2001), no. 1–2, 101–174.

[Pa59] R. E. PATTLE. Diffusion from an instantaneous point source with a concentration-dependent coefficient, *Quart. J. Mech. Appl. Math.* **12** (1959), 407–409.

[Pz81] A. PAZY. The Lyapunov method for semigroups of nonlinear contractions in Banach spaces. *J. Analyse Math,* **40** (1981), 239–262.

[Pe81] L. A. PELETIER. The porous medium equation. In *Applications of Nonlinear Analysis in the Physical Sciences*, H. Amann, N. Bazley, and K. Kirchgässner. eds., Pitmann, London (1981), 229–241.

[PZ95] M. A. PELETIER, H. ZHANG. Self-similar solutions of a fast diffusion equation that do not conserve mass, *Diff. Int. Eqns.* **8** (1995) 2045–2064.

[Per03] G. PERELMAN. *The entropy formula for the Ricci flow and its geometric applications*, http://arxiv.org/ 2002, *Ricci flow with surgery on three-manifolds*, http://arxiv.org/ 2003, *Finite extinction time for the solutions to the Ricci flow on certain three-manifolds*, http://arxiv.org/ 2003.

[Pk91] L. PERKO. *Differential Equations and Dynamical Systems.* Third edition. Texts in Applied Mathematics, **7**. Springer-Verlag, New York, 1991, revised edition 2001.

[PM90] P. PERONA, J. MALIK. Scale space and edge detection using anisotropic diffusion, *IEEE Trans. Pattern Anal. Machine Intell.* **12** (1990), 629–639.

[Pi82] M. PIERRE. Uniqueness of the solutions of $u_t - \Delta\phi(u) = 0$ with initial datum a measure, *Nonlinear Anal., TMA* **6** (1982), 175–187.

[Pi87] M. PIERRE. Nonlinear fast diffusion with measures as data. In *Nonlinear Parabolic Equations: Qualitative Properties of Solutions* (Rome, 1985), 179–188, Pitman Res. Notes Math. Ser., **149**, Longman, Harlow, 1987.

[PV93] M. M. PORZIO, V. VESPRI. Hölder estimates for local solutions of some doubly nonlinear degenerate parabolic eqns, *J. Diff. Eqns* **103** (1993), 146–178.

[QW99] Y. W. QI, M. WANG. The global existence and finite time extinction of a quasilinear parabolic equation, *Advances Diff. Eqns* (1999) **4**, 731–753.

[Rav70] P.-A. RAVIART. Sur la résolution de certaines équations paraboliques non linéaires, *J. Functional Analysis* **5** (1970), 299–328.

[ReV05] G. REYES, J. L. VÁZQUEZ. Symmetrization for weighted nonlinear elliptic and parabolic equations, *J. Eur. Math. Soc.*, **8** (2006), 531–554.

[REV97] A. RODRIGUEZ, J. L. VÁZQUEZ, J. R. ESTEBAN. The maximal solution of the logarithmic fast diffusion equation in two space dimensions, *Advances Diff. Eqns.* **2**, 6 (1997), 867–894.

[RV90] A. RODRIGUEZ, J. L. VÁZQUEZ. A well-posed problem in singular Fickian diffusion, *Archive Rat. Mech. Anal.* **110**, 2 (1990), 141–163.

[RV95] A. RODRIGUEZ, J. L. VÁZQUEZ. Non-uniqueness of solutions of nonlinear heat equations of fast diffusion type, *Ann. Inst. Henri Poincaré, Analyse Non Lin.* **12**, 2 (1995), 173–200.

[RV02] A. RODRÍGUEZ, J. L. VÁZQUEZ. Obstructions to existence in fast-diffusion equations, *J. Diff. Eqns.* **184** (2002), 348–385.

[Ro79] G. ROSEN. Nonlinear heat conduction in solid H_2, *Phys. Rev. B* **19** (1979), 2398–2399.

[Sb62] E. S. SABININA. On a class of nonlinear degenerate parabolic equations, *Dokl. Akad. Nauk SSSR* **143** (1962), 794–797, *Soviet Math. Dokl.* **3** (1962), 495–498.

[Sb65] E. S. SABININA. On a class of quasilinear parabolic equations not solvable with respect to the time derivative, *Sibirskii Mat. Zh.* **6** (1965), 1074–1100.

[Sk89] P. E. SACKS. A singular limit problem for the porous medium equation, *J. Math. Anal. Appl.* **140** (1989), no. 2, 456–466.

[SC01] L. SALOFF-COSTE. *Aspects of Sobolev-Type Inequalities*, London Mathematical Society Lecture Note Series No. 289, 2001.

[SV05] F. SALVARANI, J. L. VÁZQUEZ. The diffusive limit for Carleman-type kinetic models, *Nonlinearity* **18** (2005), no. 3, 1223–1248.

[S4-87] A.A. SAMARSKII, V. A. GALAKTIONOV, S. P. KURDYUMOV Y A. P. MIKHAILOV. *Blow-up in Problems for Quasilinear Parabolic Equations*, Nauka, Moscow, 1987 (in Russian). English transl.: Walter de Gruyter, Berlin, 1995.

[St69] D. H. SATTINGER. On the total variation of solutions of parabolic equations, *Math. Ann.* **183** (1969), 78–92.

[Sc84] R. SCHOEN. Conformal deformation of a Riemannian metric to constant scalar curvature, *J. Diff. Geometry* **20** (1984), 479–495.

[SY94] R. SCHOEN, S.-T. YAU. *Lectures on Differential Geometry,* Lecture notes prepared by W. Y. Ding, G. Q. Zhang, J. Q. Zhong and Y. Ch. Xu. *Conference Proceedings and Lecture Notes in Geometry and Topology*, I. International Press, Cambridge, MA, 1994.

[SV96] S. I. SHMAREV, J. L. VÁZQUEZ. The regularity of solutions of reaction–diffusion equations via Lagrangian coordinates, *Nonlinear Differential Eqns Appl.* **3** (1996), no. 4, 465–497.

[Sk98] I. I. SKRYPNIK. The Harnack inequality for nonlinear degenerate parabolic equations with measurable coefficients, *Dopov. Nats. Akad. Nauk Ukr. Mat. Prirodozn. Tekh. Nauki* (1) (1998) 57–59.

[Sb38] S. L. SOBOLEV. On a theorem in functional analysis, *Mat. Sbornik* **4** (1938), pp. 471–497 (in Russian; English translation Am. Math. Soc. Translations, **2 (34)** (1963), pp. 39–68).

[Ta76] G. TALENTI. Best constant in Sobolev inequality, *Ann. Mat. Pura Appl.* (4) **110** (1976), 353–372.

[Ta76b] G. TALENTI. Elliptic equations and rearrangements, *Ann. Scuola Norm. Sup.* (4) **3** (1976), 697–718.

[Ta79] G. TALENTI. Nonlinear elliptic equations, rearrangements of functions and Orlicz spaces, *Annal. Mat. Pura Appl.* 4, **120** (1979), 159–184.

[Ta93] G. TALENTI. On the decay of solutions to parabolic PDE. In *Boundary Value Problems for Partial Differential Equations and Applications*, 429–435, RMA Res. Notes Appl. Math., **29**, Masson, Paris, 1993.

[Tr68] N. S. TRUDINGER. On Harnack type inequalities and their applications to quasi linear parabolic equations, *Comm. Pure. Appl. Math.* **21** (1968), 205–226.

[U86] M. UGHI, A degenerate parabolic equation modelling the spread of an epidemic, *Ann. Mat. Pura Appl.* (4) **143** (1986), 385–400.

[Va82] J. L. VÁZQUEZ. Symétrisation pour $u_t = \Delta\varphi(u)$ et applications, *C. R. Acad. Sc. Paris* **295** (1982), 71–74.

[Va82b] J. L. VÁZQUEZ. Symmetrization in nonlinear parabolic equations, *Portugaliae Math.* **41** (1982), 339–346.

[Va84] J. L. VÁZQUEZ. The interfaces of one-dimensional flows in porous media, *Trans. Amer. Math. Soc.* **285**, no. 2 (1984), 717–737.

[Va90] J. L. VÁZQUEZ. Two nonlinear diffusion equations with finite speed of propagation, *Problems Involving Change of Phase*, K. Kirchgässner ed., Springer Lecture Notes in Phys. **359**, Springer Verlag, Berlin, 1990, pp. 197–206.

[Va92] J. L. VÁZQUEZ. Nonexistence of solutions for heat nonlinear equations of fast-diffusion type, *Journal Math. Pures et Appliquées*, **71** (1992), 503–526.

[V92b] J. L. VÁZQUEZ. An introduction to the mathematical theory of the porous medium equation. In *Shape Optimization and Free Boundaries*, M. C. Delfour ed., Math. and Phys. Sciences, Series C, Kluwer Academic, Dordrecht, 1992 pages 347–389.

[Va96] J. L. VÁZQUEZ. Entropy solutions and the uniqueness problem for nonlinear second-order elliptic equations, in *Nonlinear Partial Differential Equations,* A. Ben Kirane and J.P. Gossez eds., Pitman Research Notes in Mathematics Series **343**, Addison-Wesley Longman, Harlow, 1996, pages 179–203.

[Va03] J. L. VÁZQUEZ. Asymptotic behaviour for the porous medium equation posed in the whole space, *J. Evol. Eqns* **3** (2003), 67–118.

[Va04] J. L. VÁZQUEZ. Asymptotic behaviour for the PME in a bounded domain. The Dirichlet problem, *Monatshefte für Mathematik*, **142**, nos 1-2 (2004), 81–111.

[Va04b] J. L. VÁZQUEZ. Symmetrization and mass comparison for degenerate nonlinear parabolic and related elliptic equations, *Advanced Nonlinear Studies* **5** (2005), 87–131.

[Va04c] J. L. VÁZQUEZ. Darcy's Law and the theory of shrinking solutions of fast diffusion equations, *SIAM J. Math Anal.* **35**, no 4 (2004), 1005–1028.

[Va05] J. L. VÁZQUEZ. Failure of the strong maximum principle in nonlinear diffusion. Existence of needles, *Comm. Partial Diff. Eqns.* 30, no 9 (2005), 1263–1303.

V05b] J. L. VÁZQUEZ. Evolution of point masses by planar logarithmic diffusion, *preprint,* 2005.

V06] J. L. VÁZQUEZ. *The Porous Medium Equation. Mathematical theory*, Oxford University Press, to appear.

[VER96] J. L. VÁZQUEZ, J. R. ESTEBAN, A. RODRIGUEZ. The fast diffusion equation with logarithmic nonlinearity and the evolution of conformal metrics in the plane, *Advances Diff. Eqns* **1**, no. 1 (1996), 21–50.

[Ve79] L. VÉRON. Effets régularisants des semi-groupes non linéaires dans des espaces de Banach, *Ann. Fac. Sci. Toulouse Math.* (5) **1**, no. 2 (1979), 171–200 (in French).

[Wol] E. W. WEISSTEIN. Gauss–Bonnet Formula. From MathWorld–A Wolfram Web Resource. http://mathworld.wolfram.com/Gauss-BonnetFormula.html

[Wu93] L. F. WU. The Ricci flow on complete R^2, *Comm. Anal. Geom.* **1**, no. 3–4 (1993), 439–472.

[Wu93b] L. F. WU. A new result for the porous medium equations derived from the Ricci flow, *Bull. Amer. Math. Soc.* **28** (1993), 90–94.

[WZYL] ZH. Q. WU, J. N. ZHAO, J. X. YIN, H. L. LI. *Nonlinear Diffusion Equations*, World Scientific, Singapore, 2001.

[Ya60] H. YAMABE. On a deformation of Riemannian structures on compact manifolds, *Osaka Math. J.* **12** (1960), 21–37.

[Ye94] R. YE. Global existence and convergence of Yamabe Flow, *J. Diff. Geom.* **39** (1994), 35–50.

[ZK50] YA. B. ZEL'DOVICH, A. S. KOMPANEETS. Towards a theory of heat conduction with thermal conductivity depending on the temperature, in *Collection of Papers dedicated to 70th Anniversary of A. F. Ioffe, Izd. Akad. Nauk SSSR*, Moscow, 1950, 61–72.

[ZR66] YA. B. ZEL'DOVICH, YU. P. RAIZER. *Physics of Shock Waves and High-Temperature Hydrodynamic Phenomena* II, Academic Press, New York, 1966.

[Zh95] J. N. ZHAO. The Cauchy problem for $u_t = \mathrm{div}(|\nabla u|^{p-2}\nabla u)$ when $2N/(N+1) < p < 2$, *Nonlinear Anal.* **24**, no. 5 (1995), 615–630.

Index

Lightning Source UK Ltd.
Milton Keynes UK
UKOW06n2232061215

264097UK00010B/195/P

9 780199 202973